T0327375

Advances in
Solid Oxide Fuel Cells IV

Advances in Solid Oxide Fuel Cells IV

A Collection of Papers Presented at the 32nd International Conference on Advanced Ceramics and Composites January 27–February 1, 2008 Daytona Beach, Florida

Editors

Prabhakar Singh
Narottam P. Bansal

Volume Editors

Tatsuki Ohji
Andrew Wereszczak

The American Ceramic Society

WILEY

A John Wiley & Sons, Inc., Publication

Published by John Wiley & Sons, Inc., Hoboken, New Jersey.
Published simultaneously in Canada.

For general information on our other products and services or for technical support, please contact our
Customer Care Department within the United States at (800) 762-2974, outside the United States at
(317) 572-3993 or fax (317) 572-4002.

Wiley also publishes its books in a variety of electronic formats. Some content that appears in print may
not be available in electronic format. For information about Wiley products, visit our web site at
www.wiley.com.

Library of Congress Cataloging-in-Publication Data is available.

ISBN 978-0-470-34496-5

10 9 8 7 6 5 4 3 2 1

Contents

FABRICATION

CHARACTERIZATION AND TESTING

ELECTRODES

SEALS

ELECTROLYZER

Preface

The Fifth International Symposium on Solid Oxide Fuel Cells (SOFC): Materials, Science, and Technology was held during the 32nd International Conference and Exposition on Advanced Ceramics and Composites in Daytona Beach, FL, January 27 to February 1, 2008. This symposium provided an international forum for scientists, engineers, and technologists to discuss and exchange state-of-the-art ideas, information, and technology on various aspects of solid oxide fuel cells. A total of 120 papers were presented in the form of oral and poster presentations indicating strong interest in the scientifically and technologically important field of solid oxide fuel cells. Authors from 17 countries (Australia, Austria, Brazil, Canada, China, Denmark, Germany, India, Italy, Japan, Netherlands, South Korea, Switzerland, Taiwan, Ukraine, United Kingdom, and U.S.A.) participated. The speakers represented universities, industries, and government research laboratories.

These proceedings contain contributions on various aspects of solid oxide fuel cells that were discussed at the symposium. Twenty five papers describing the current status of solid oxide fuel cells technology and the latest developments in the areas of fabrication, characterization, testing, performance, long term stability, anodes, cathodes, electrolytes, interconnects, seals, cell and stack design, proton conductors, electrolyzer, etc. are included in this volume. Each manuscript was peer-reviewed using The American Ceramic Society review process.

The editors wish to extend their gratitude and appreciation to all the authors for their contributions and cooperation, to all the participants and session chairs for their time and efforts, and to all the reviewers for their useful comments and suggestions. Financial support from the American Ceramic Society is gratefully acknowledged. Thanks are due to the staff of the meetings and publications departments of The American Ceramic Society for their invaluable assistance. Advice, help and cooperation of the members of the symposium's international organizing committee (Tatsumi Ishihara, Tatsuya Kawada, Nguyen Minh, Mogens Mogensen, Nigel Sammes, Robert Steinberger-Wilkens, Jeffry Stevenson, and Eric Wachsman) at various stages were instrumental in making this symposium a great success.

It is our earnest hope that this volume will serve as a valuable reference for the engineers, scientists, researchers and others interested in the materials, science and technology of solid oxide fuel cells.

Prabhakar Singh
Pacific Northwest National Laboratory

Narottam P. Bansal
NASA Glenn Research Center

Introduction

Organized by the Engineering Ceramics Division (ECD) in conjunction with the Basic Science Division (BSD) of The American Ceramic Society (ACerS), the 32nd International Conference on Advanced Ceramics and Composites (ICACC) was held on January 27 to February 1, 2008, in Daytona Beach, Florida. 2008 was the second year that the meeting venue changed from Cocoa Beach, where ICACC was originated in January 1977 and was fostered to establish a meeting that is today the most preeminent international conference on advanced ceramics and composites

The 32nd ICACC hosted 1,247 attendees from 40 countries and 724 presentations on topics ranging from ceramic nanomaterials to structural reliability of ceramic components, demonstrating the linkage between materials science developments at the atomic level and macro level structural applications. The conference was organized into the following symposia and focused sessions:

Symposium 1	Mechanical Behavior and Structural Design of Monolithic and Composite Ceramics
Symposium 2	Advanced Ceramic Coatings for Structural, Environmental, and Functional Applications
Symposium 3	5th International Symposium on Solid Oxide Fuel Cells (SOFC): Materials, Science, and Technology
Symposium 4	Ceramic Armor
Symposium 5	Next Generation Bioceramics
Symposium 6	2nd International Symposium on Thermoelectric Materials for Power Conversion Applications
Symposium 7	2nd International Symposium on Nanostructured Materials and Nanotechnology: Development and Applications
Symposium 8	Advanced Processing & Manufacturing Technologies for Structural & Multifunctional Materials and Systems (APMT): An International Symposium in Honor of Prof. Yoshinari Miyamoto
Symposium 9	Porous Ceramics: Novel Developments and Applications

Symposium 10	Basic Science of Multifunctional Ceramics
Symposium 11	Science of Ceramic Interfaces: An International Symposium Memorializing Dr. Rowland M. Cannon
Focused Session 1	Geopolymers
Focused Session 2	Materials for Solid State Lighting

Peer reviewed papers were divided into nine issues of the 2008 Ceramic Engineering & Science Proceedings (CESP); Volume 29, Issues 2-10, as outlined below:

- Mechanical Properties and Processing of Ceramic Binary, Ternary and Composite Systems, Vol. 29, Is 2 (includes papers from symposium 1)
- Corrosion, Wear, Fatigue, and Reliability of Ceramics, Vol. 29, Is 3 (includes papers from symposium 1)
- Advanced Ceramic Coatings and Interfaces III, Vol. 29, Is 4 (includes papers from symposium 2)
- Advances in Solid Oxide Fuel Cells IV, Vol. 29, Is 5 (includes papers from symposium 3)
- Advances in Ceramic Armor IV, Vol. 29, Is 6 (includes papers from symposium 4)
- Advances in Bioceramics and Porous Ceramics, Vol. 29, Is 7 (includes papers from symposia 5 and 9)
- Nanostructured Materials and Nanotechnology II, Vol. 29, Is 8 (includes papers from symposium 7)
- Advanced Processing and Manufacturing Technologies for Structural and Multifunctional Materials II, Vol. 29, Is 9 (includes papers from symposium 8)
- Developments in Strategic Materials, Vol. 29, Is 10 (includes papers from symposia 6, 10, and 11, and focused sessions 1 and 2)

The organization of the Daytona Beach meeting and the publication of these proceedings were possible thanks to the professional staff of ACerS and the tireless dedication of many ECD and BSD members. We would especially like to express our sincere thanks to the symposia organizers, session chairs, presenters and conference attendees, for their efforts and enthusiastic participation in the vibrant and cutting-edge conference.

ACerS and the ECD invite you to attend the 33rd International Conference on Advanced Ceramics and Composites (http://www.ceramics.org/daytona2009) January 18–23, 2009 in Daytona Beach, Florida.

TATSUKI OHJI and ANDREW A. WERESZCZAK, Volume Editors
July 2008

Technical Overview

RESEARCH ACTIVITIES AND PROGRESS ON SOLID OXIDE FUEL CELLS AT USTC

Guangyao Meng, Ranran Peng, Changrong Xia, Xingqin Liu

USTC Laboratory for Solid State Chemistry and Inorganic Membranes
Department of Materials Science and Engineering, University of Science and Technology of China
(USTC), Hefei 230026, China ⎵mgym@ustc.edu.cn⎵

ABSTRACT
    This article briefly introduces the research activities and progress on Solid Oxide Fuel Cells (SOFC) at USTC. Lab. directed by one of the present author, prof. Meng, in recent ten years. The content includes the following topics:
    (1) Searching new electrolyte materials for SOFCs
    (2) Development of preparation techniques for thin electrolyte membranes on porous anode support
    (3) Modification of both cathode and anode by nano-techniques
    (4) Tubular CMFCs : low cost fabrication and ceramic interconnect materials
    (5) Ammonia fueled CMFCs with proton and oxide ion electrolytes

INTRODUCTION
    Solid Oxide Fuel Cells (SOFCs) have attracted worldwide interest for their high energy conversion efficiency, structure integrity, easy operation, and less compact to environment as well as the high tolerance to fuels.
    The research on SOFCs was started relatively late in China, and in late 1980's, there were only a few research groups dealing with materials related to SOFCs. The major event for R & D of SOFCs in China was the 97th Xiangshan Scientific Conference, topic of which was 'New Solid Fuel Cells', held in June 14th – 17th, 1998, Xiangshan Hotel, Beijing. The conference established developing the intermediate temperature SOFCs (IT-SOFCs) as the main target in R & D of SOFCs, and the routes of searching high performance key materials, developing techniques to prepare thin electrolyte membranes on porous electrode supports as well as preparing active electrodes with nano-microstructures were proposed in order to realize IT- SOFCs [1]. Since then, laboratory for solid state chemistry and inorganic membranes at USTC as one of the major units dealing with SOFCs has joined in the main research projects on SOFCs granted by NSFC and MSTC in the 10th five year program in China. In recently years, our work emphasis has been put on the R and D of tubular ceramic membrane fuel cells (CMFCs) from viewpoint of practical applications and following a research route based on ˙counter-main stream consideration'. This article would introduce briefly the results in these research activities in the following sections.

1. Searching and investigation of new electrolytes rather than yttrium stabilized zirconia (YSZ)
    In order to pursue IT-SOFCs, early research work was first focused on searching new electrolyte materials to substitute YSZ for its rather low conductivity particularly at temperature below 800°C. In a work trying to utilize high proton conductive Li_2SO_4 in the dual phase of $Li_2SO_4 \llcorner Al_2O_3 \llcorner + Ag$ as a hydrogen permeation reactor, we noticed the H_2S formation in H_2 which was further confirmed in a

3

H_2/O_2 cell with Li_2SO_4 Al_2O_3 as electrolyte[2] due to the following reaction occurred,

$$Li_2SO_4 \quad 4H_2 \rightarrow H_2S\uparrow \quad 2LiOH \quad 2H_2O$$

This work reminded researchers the importance to consider the thermodynamic stability of the electrolyte materials that had not been paid sufficient attention. Knowing chloride exhibiting high chemical stability Dr. S.W. Tao who was a Ph.D. student of mine then made an attempt to use doped NaCl (adding 70% Al_2O_3 to enhance the strength) as electrolyte to assembly H_2 and O_2 concentration cells and found the remarkable O^{2-}/H^+ conduction in 650-750°C with oxide ion transference number of 0.98 at 700°C [3]. Further investigation found that a composite of $LiCl$ $SrCl_2$ with doped ceria exhibited even higher conductivity at the temperature above the eutectic point 485°C for 53mol LiCl _ 47mol_ $SrCl_2$. As can be seen from the V-I and P-I curves of a cell consisted of Ni-GDC anode, $LiCl$-$SrCl_2$–GDC (Gd doped CeO_2) electrolyte (0.40mm thick) and $LiNiO_2$ cathode, shown in Fig.1, the open circuit voltages of the cell (OCV) are close to 1.2V indicating the pure ionic conductivity of the electrolyte material and the peak power densities of the cell are in the range of 120 – 270 mW/cm^2 in 460 _ 550°C [4]. And another cell showed the even better performance with a peak power density of 500 mW/cm^2 at 625°C [5]. The data were remarkably higher than the best record at that time, 140 mW/cm^2 at 500°C by Doshi et. al., with the cell based on GDC electrolyte about 30 µm in thickness [7].

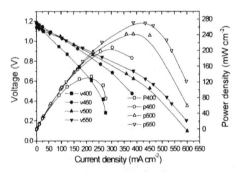

Fig. 1 V-I and P-I curves of a cell consisted of Ni-GDC anode, GDC-LiCl-SrCl$_2$ electrolyte (0.40mm thick), and LiNiO$_2$ cathode [4,5]

The electrolyte conductivity versus temperature were roughly obtained from the slope of the cell V-I curves and shown in Fig. 2[6]. It can be seen that the conductivity of the composite electrolyte is about 2~10 times higher than that of pure GDC or LSGM ($La_{0.6}Sr_{0.2}Ga_{0.8}Mg_{0.2}O_{3-\delta}$), and 1~2 orders of magnitude higher than that of YSZ in the temperature range of 400~600°C. And the most interesting characteristics was that the conductivity was not only high but also the activation energy of the conduction was quite low and less sensitive to the temperature, compared with all the well known oxide ion electrolytes. This was most attractive for the development of IT-SOFCs. To interpret the high conductivity with no record before, a model for the electric conduction mechanisms was proposed [8]. The model supposed that there were four possible paths for the electric charge carriers to go through: (1) continuous molten chloride salt, (2) continuous ceria particles, (3) continuous ceria- molten salt and (4) disconnected ceria particle- molten salt ambient. Possibly, the Path (2) is the most conductive path because the molten salt exhibits much higher ionic mobility and the path of GDC-Chlorides interface is also the easy way to go for ions. In the case of continuous solid GDC particles, the cell OCV may lower than 0.9V due to the partial electronic conduction of GDC. But the cell could have higher power density because of more efficient parallel ionic paths that was proved to be true [6,8,9].

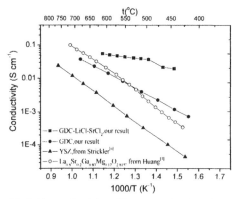

Fig.2 The conductivity of the composite electrolyte

The conductive salt-oxide composites demonstrated surely an attractive new route to search more efficient electrolyte systems with unique characteristics for reduced temperature fuel cells. After further investigation, however, we discovered that the cells with these composite electrolytes could not keep long duration due to the volatility of the salt component, particularly in the gas flow systems. The study on such material systems was stopped for many years, but we do think this kind of electrolyte systems may find their proper usages in future.

As to the well known alternative oxide conductors, including doped $LaGaO_3$ and doped CeO_2 (GDC or SDC), our investigation was mainly put on developing so called 'soft chemical synthesis' routes to prepare high reactivity powders and optimizing the properties by composition refinement[10-19]. For the $La_{0.9}Sr_{0.1}Ga_{0.8}Mg_{0.2}O_3$(LSGM) the powder prepared by citrate method reached a 97% relative density at 1450°C [10] while the densification temperature was usually around 1600°C for the powders by conventional solid phase reaction. With $La_{0.6}Sr_{0.4}Ga_{0.8}Ni_{0.2}O_{3-\delta}$ as a compatible anode, the cell with LSGM electrolyte of 0.5mm provide a power density of 270 mW/cm^2 at 750°C, predicting the even much higher performance for the cells with thinner electrolyte [11]. Owing to the less Ga source and lack of compatible electrode materials we turned the research efforts onto GDC and SDC for IT-SOFCs [12-19]. Our investigation showed that Sm doped CeO_2, $Ce_{0.8}Sm_{0.2}O_{2-\delta}$ or $Ce_{0.85}Sm_{0.15}O_{2-\delta}$, exhibited better properties than GDC which got more reports in the literature. As shown in Fig. 3, the OCV value of the cell with SDC can be above than 0.9 V when operates at a temperature lower than 700°C [13]. The SDC powder prepared by a polyvinyl alcohol-induced low temperature synthesis had a particle size of 20-30 nm and could be densified into 98 relative density at 1300°C and got a conductivity of 0.033S/cm at700°C that was a quite good value.[14]. And based on the nano-particle powders prepared by such a polymer assisted process, a method so called triple layer co-pressing and co firing was developed in our laboratory to make disc fuel cells with very thin electrolyte membrane which has been the powerful route to investigate materials and cell performance

[19]. Fig. 4 is the result from such a cell with thin membrane electrolyte [20]. We may see that the SDC membrane is only about 12 μm and the power density reaches 1872 mW/cm^2 at 650°C and 748 mW cm^{-2}at 550°C, which are surely in the highest range in literature. The co-pressing and firing process, however, is only suitable for laboratory but not for scaling up and for the later it will be described in the nest section.

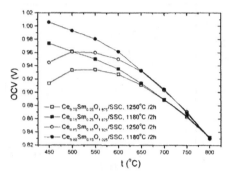

Fig. 3   OCVs of the cells with SDC electrolyte: the effect of interface microstructure on OCVs (electrode sintering condition) [13]

Fig. 4 The microstructure(A) and performance(B) of a cell made by co-pressing and co-firing Ni+SDC anode and BSCF cathode with SDC electrolyte the powders made by polymer assisted combustion method [20]

2. Fabrication techniques for thin PEN membranes on porous anode supports

It is of essential significance to develop proper techniques to fabricate thin electrolyte membrane on porous anode support for reducing operation temperature and enhancing performance of SOFCs with ether YSZ electrolyte and other high conductive electrolytes. And it has been commonly recognized that the advanced ceramic processing and co-firing of multi-layers would be the right route as low cost fabrication techniques for SOFCs. A number of techniques, usually the polymer assisted ceramic processing was developed to make dense thin layer of electrolyte on porous anode of YSZ + NiO or DCO + NiO and then deposit a porous cathode. As the results from the cells summarized in Table 1, the techniques were all successful to make a thin and dense electrolyte as thin as 10 to 50 μm in our attempts [21-26].The tape casting technique was readily employed to make both support and the top electrolyte layer, by which the bi-layers fabricated were good at co-firing, but the cell performance was not so satisfied [21]. The silk screen printing was the first process for us to successfully prepare thin $Ce_{0.8}Y_{0.2}O_{1.9}$ electrolyte of 15μm and got a pretty high cell power density of 360 mW/cm^2 at 650°C [22]. Multi-Spin-coating technique could provide very thin electrolyte and got a fairly high cell performance even with YSZ electrolyte [23], but is not suitable to calling up and also not cost effective. The modified dip-coating and powder spray process are of cost effective and suitable to scaling up for both planner and tubular SOFCs [24-26].

Table 1 Various techniques for thin electrolyte membranes and the cell performance, developed at USTC Lab. of SSC & IM.

Technique for thin membranes	Electrolyte material	Sintering temperature (°C)	Thickness (μm)	Power density (mW/cm^2)	Refer-ence
Tape casting	SDC/NiO -SDC	1400	50	260 (700°C)	[21]
Screen printing	$Ce_{0.8}Y_{0.2}O_{1.9}$	1350	15	360 (650°C)	[22]
Spin-coating	YSZ/Ni-SDC	1300	12	535 (750°C)	[23]
Dip- coating	YSZ	1400	30	190 (800°C)	[24]
Electrostatic spray	YSZ	1400	15	315 (800°C)	[25]
Suspension spray	YSZ	1400	10	837 (800°C)	[26]

Suspension spray technique is not only the right technique to fabricate electrolyte membranes on porous anode or cathode support as thin as around 10μm, but also the right process to make active or transition layers to modify the electrode interfaces[24-26]. Fig. 5 shows the result of a fuel cell made by suspension spray technique on porous disk anode [26]. The YSZ electrolyte membrane was around 10μm and the maximum powder density was only 400 mW/cm^2 at 800°C probably due to the cracks and pores on the YSZ - anode interface as seen in Fig.5(a). After making a modification layer on the rough surface of the anode by the same process (see Fig. 5-b), the power density of the cell increased to 837 mW/cm^2 at 800°C, but still 214 mW/cm^2 at 650°C (Fig.5-c)|

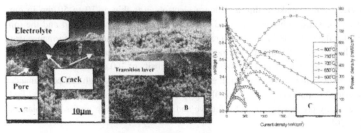

Fig.5 The microstructures and performance of the cells made by suspension spray process[26]

As seen from Fig.5 (b), the interface of YSZ and cathode is still poor, and thus there is obvious electrode polarization on the V-I curves of the cell (Fig.5-c). A recent result shows that after adding a SDC active layer by the suspension spray process on YSZ surface before coating cathode, cell power density reaches 443 mW/cm^2 at 650°C and 187 mW/cm^2 at 600°C, as seen from the Fig.6 [27]. This means that YSZ could also be used as electrolyte material for IT-SOFCs as long as the proper fabrication technique developed.

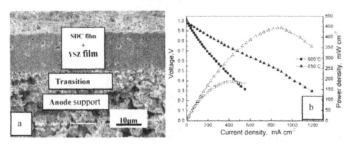

Fig 6 (a) Section view of the cell with anode transition layer and active cathode SDC layer, and (b) V-I, P-I curves of the cell in Fig.6-a

3. Modification of both cathode and anode by nano-techniques

It has been recognized that the composition and microstructure of the electrodes are the major factors to affect both performance and life duration of a cell. In recent years, our research work on electrodes has included following topics:

- New material systems[28-30]
- Interface modification of anode or cathode by chemical routes[25-27,31-33]
- New designs of electrode structure and preparation for the long term stability[34-36]

As mentioned above that modification of the interfaces by spraying nano-particles made by wet chemical routes could provide a significant enhancement to the cell performance because of the increase of triple phase boundaries and the improvement of interface coherency [25-27,31-33]. Both anode and cathode still have their own problems, such as the carbon deposit on Ni based anode when hydrocarbon fuels are used and the contradiction of catalysis activity of cathode materials and their thermal expansion consistence with electrolyte. Our lab. firstly investigated the performance of SOFCs with natural gas and biomass gas and got a peak power density of over 300mW/cm^2 at 600°C for the cell of Ni SDC/SDC/SSC with biomass fuel[34]. In order to create a high performance anode with against carbon-deposit, proper catalytic activity, structure stability as well as higher ionic conductivity a new anode structure with branch-like-microstructure was designed recently [35]. The anode consisted of porous Ni-SDC and micron size SDC to form a continuous branch like structure coated with nano-particle SDC. It gives a number of advantages:

1   Against coking on anode because of the nano-SDC coatings on Ni-SDC surface
2   High electrochemical activity comes from nano-size SDC particles which exhibit high oxidation reactivity.
3   High conductivity from Ni based Ni-SDC anode
4   Ni-SDC based anode is compatible to SDC electrolyte thermo- mechanically, thus thermodynamic stable
5   Easy to fabricate, the simple dip-coating process can be employed to coat nano-SDC on Ni - SDC anode

As presented in Fig. 7, the cells with new structure anode coated with various amount of nano-SDC particles display a great improvement in their cell performance against carbon-deposit with methane as fuel. The cell with 25 mg/cm^2 SDC coating was operated in CH$_4$ at 600°C for 50 hrs without obvious decrease in power output or structure change. While the power density of the cell without SDC coating on anode decreased by 60% after only 10hrs operation [34].

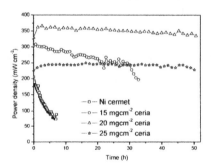

Fig.7 The longer term performance of SOFC cells with various SDC coating on Ni-SDC anode for CH$_4$ as fuel, operated at 0.5V and 600°C

Similar to the anode structure described above, the cathode side can be improved by the same idea. With nano-LSC (La$_{1-x}$Sr$_x$CoO$_3$) coating on porous and branch like cathode the cell illustrated very high performance stability in longer term and multi-thermal cycles as shown in Fig. 8[35]. As can be seen that the area specific resistance(ASR) (measured by ac impedance spectroscopy technique in situ)of a cell with a conventional SDC-LSC cathode made by silk screen printing increased obviously, from the original value of 2.4Ωcm^2 to 3.5Ωcm^2 during the thermal cycles between 500-800°C for 20 times in 20 days and further increased to 12.5 Ωcm^2 during thermal cycles of 10 times from room temperature to 800 °C in 10 days, and then remained changeless at 600°C for more than 60 days. While the ASR of the cell with new cathode coated by nano-size LSC has a very low value (0.30 Ωcm^2) and kept stable during the testing for more than 100 days. The results demonstrate solidly that the novel design of the electrodes has remarkably improved the performance of SOFCs that is certainly promising for the commercialization of this new energy source.

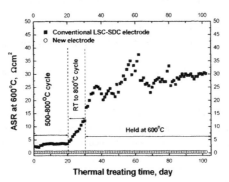

Fig.8 A comparison in ASR of the new designed cathode with conventional SDC-LSC composite cathode during thermal cycling in longer term

4. Tubular CMFCs: design, fabrication [37-41] and interconnect ceramics [42-50]

The first attempt according to 'counter-main stream consideration route' was turned on the development of tubular SOFCs. A new tubular design of anode supported with multi-gas tunnels shown in Fig. 9 was proposed and patented [36]. This configuration exhibits 3 major characteristics:

(1) The fuel (e.g. CH$_4$ + 3% H$_2$O) inlets through the central tunnel and flies out through the other tunnels, thus it can easily perform internal reforming.

(2) The cathode surface is designed in wave or tentacle form so that the effective electrode area will be increased by 40-50% compared with flat surface.

(3) It can be easily fabricated by cost effective ceramic processing techniques which are developed in the lab. , including extrusion, gel-casting[37], silk screen printing[22], dip-coating[24] and suspension spray[25,26] etc.

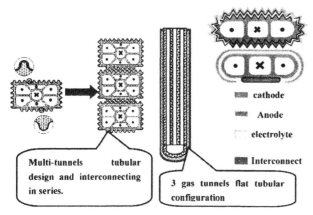

cathode

Anode

electrolyte

Interconnect

Multi-tunnels tubular design and interconnecting in series.

3 gas tunnels flat tubular configuration

Fig.9 _ China patent ZL02113198.8

The tubular anode (Ni-YSZ) supports has been fabricated by extrusion, gel-casting [37] and the techniques described above have been employed to make single PEN cells on the tubular anode support [24-26]. Fig. 10 shows the morphology of a small round tubular single cell and its performance [40]. The cell power density was improved very much, when the interface modification was made on anode and cathode [38]. As we can see that the peak power density(Fig.10-A) of the cell with YSZ electrolyte membrane in 20μm is over 400 mW/cm^2 at 850°C and gradually decreased to 270 mW/cm^2 at 700°C, indicating the smaller ASR contribution to the total cell resistance[41]. The SEM picture of the cell section (Fig.10-B) shows very intimately electrode interfaces that display the better cell performance.

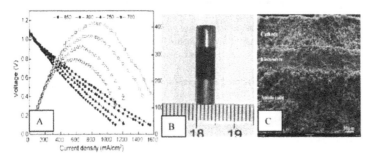

Fig 10 the cell performance and microstructure of a tubular cell made by cost effective process [40, 41]
    (A) V-I, P-I curves of a tubular Cell fueled with H$_2$
    (B) Picture of the cell made by dip-coating, and
    (C) Section view of the cell, showing the very well coherent electrode interfaces.

For planner SOFCs, the metal based materials could be chosen to make interconnect. But for the stacks of tubular cells, the interconnect layer must be ceramic and directly prepared on the tubes. Doped chromates, typically $La_{0.7}Ca_{0.3}Cr_3$ (LCC) and $La_{0.7}Sr_{0.3}CrO_3$ (LSC), exhibit excellent properties, particular high stability in both oxidant and reducing ambient. But two major shortcomings: lower electric conductivity and too high temperature for densification, hinter its applications, especially for cost effective fabrication of IT-SOFCs. To realize the tubular cell stacks we have done much effort to search new material systems and obtained progress [42-50]:

1) Full or partial substitution of La in $La_{0.7}Ca_{0.3}CrO_3$ LCC , the best interconnect ceramics, by other rare earth element (Gd, Pr Nd Tb) much increased the conductivity of the materials. For instance, the conductivity of $Gd_{0.7}Ca_{0.3}CrO_3$ GCC at 700°C in air was 24 S/cm 30% higher than LCC 18.5 S/cm and 130% higher than LSC 10.4 S/cm . The more important is that its conductivities in H_2 are 8.4 S/cm at 900°C and 6 S/cm at 500°C, which are much higher than that reported in literature, e.g. 1.5/cm at 900°C for $La_{0.75}Sr_{0.25}Cr_{0.5}Mn_{0.5}O_3$ reported by Tao et al[51].

2) It was found that doping DCO(GDC, SDC, YDC) into LCC created a new structure or form a composite, which displayed extremely high conductivity[43-47]. As listed in Table 2, the conductivity values for samples of LCC doped with 2-10 % SDC were 5-38 time higher than LCC and the oxygen permeation and thermal expansion coefficient remain changeless compared with LCC. Another interesting point was that the samples showed a relative density of 97-98%, indicating that the sintering ability of the materials was also much improved. This was probably related to the nano-size particle prepared by the soft chemistry method [42].

3) More recently, the sintering temperature for densification has been further lowered by putting sintering adds[48,49] and controlling B site deficiency[50]. The results showed that Zn doped LCC sintered at 1250 1400 C for 5hrs could obtain 96 98 relative density and its conductivity reached 45.7 S cm^{-1} at 800°C and 34.5 S cm^{-1} at 500°C in air, and 2.06-6.1 S cm^{-1}in H_2, respectively.

These results illustrated that the new material systems have resolved the two major problems of the conventional LCC or LSC, and can well meet the requirements for tubular SOFC stacks. Particularly, the ceramic interconnect can be fabricated by utilizing low cost ceramic processing and co-firing at 1350 1400°C .

Table 2. Properties of LCC with adding SDC as new interconnect ceramics [45]

SDC content (weight%)	Relative density (%)	TEC at 1000°C (10^{-6} K^{-1})	The oxygen permeation flux/mol S cm^{-1}	Electrical conductivity σ at 800°C (Scm^{-1}) in air
0	97 4	11.12	6.51×10^{-9}	17.77
2	97 6	11.33	9.76×10^{-9}	180.25
4	97 7	11.65	2.33×10^{-9}	341.76
5	97 9	11.93	4.69×10^{-9}	687.81
6	98 1	12.22	5.23×10^{-9}	127.72
8	98 4	12.24	6.91×10^{-9}	129.09
10	98 7	12.46	7.88×10^{-9}	96.18

5. Ammonia fueled CMFCs with proton or oxide ion electrolytes

Industrial Liquid ammonia directly fueled CMFCs has been one of the research activities recent years at USTC Lab. [52-57], initiated by "counter main-stream consideration in R & D of SOFCs", noticing that great efforts have been made to search ways to prepare pure H_2 for PEMFC and to resolve the anode coking problem for SOFCs. Nowadays exploring proper fuels seems to be crucial for the commercialization of SOFCs since there are significant difficulties for pure hydrogen and hydrocarbons, the now-extensively used fuels for SOFCs. Pure hydrogen is both expensive and hard to store or transport; hydrocarbons will cause a severe coking for traditional Ni anode of SOFCs, and little progress has been made to find replacements for Ni. Ammonia is a good hydrogen carrier and a less concerned feedstock for SOFCs, and will be a nice substitute for hydrogen and hydrocarbons in fuel cells for the following reasons:

- High energy density. It contains 75 mol % H and the volumetric energy density of liquid ammonia is about 9×10^6 kJm^{-3}, higher than that of liquid hydrogen.

- Relative safe. Ammonia is less flammable compared with other fuels and the leakage of ammonia can be easily detected by the human nose under 1 ppm.

- Great suitability to CMFCs-O or CMFCs-H [55]. There are no concerns about anode coking and un-stability of $BaCeO_3$ based proton electrolytes due to the CO_2 existence, because of no carbon species in anode apartment and in case of CMFC-H cases the H_2O formed at cathode-electrolyte interface would hinder the diffusion of CO_2 possibly existed in air as oxidant to electrolyte.

- The right candidate of liquid fuel for distributed and portable SOFC/CMFCs devices, at least at the present stage when the coking problem of hydrocarbon fuels is not yet resolved.

- Low price and good competition for CMFC marketing. The price of ammonia is as competitive as hydrocarbons, 30-40 % of LPG and petroleum.

Attempt to use NH_3 as fuel for YSZ based SOFC was first made by Wojcika et al. [58], but was paid little attention, probably because of the lower cell performance(about 50 mW cm^{-2} at 800°C due to the thick YSZ electrolyte supported cell with Pt as electrode) and the worry that the toxic NOx may be produced:

$$2NH_3 + 5O^{2-} = 2NO + 3H_2O + 10e^-$$

Our first work was based on a full ceramic cell of anode supported thin proton electrolyte($BaCe_{0.8}Gd_{0.2}O_{2.9}$, 50μm thick) and achieved a maximum power density of 355 mW cm^{-2} at 700°C, which was in the range accepted for applications. For comparison, cells were also tested at 700°C with hydrogen as fuel, where the power density was about 371mW/cm^2 [52]. The subsequent research works were trying to answer the interested problems, such as NH_3 usage efficiency, the possible NOx formation in case of oxide ion electrolyte cell (CMFC-O) as well as the performance with different electrolytes in various thickness [53-57]. The results have been fairly positive and attractive, which are summarized bellow:

(1) As theoretically expected, the Ni based anode was the effective catalysis for NH_3 thermal decomposition, and conversion of NH_3 into H_2 and N_2 was experimentally determined to be completed (> 99%) above 500°C, depending on the gas flow rate. The smaller the gas flow rate, the more completely the ammonia decomposes. [53]

(2) It is proved by from the experimental OCV data of the cells that it is the H_2 instead of NH_3 itself responsible for anode process in both CMFC-H and CMFC-O [52-57].

(3) For CMFC-O fueled by NH_3, there was no NOx detected and it was consistent to the theoretical prediction that on anode it is O^{2-} ions from cathode, which exhibit much less oxidative reactivity than oxygen molecules or atoms [54].

(4) For CMFCs with a thicker electrolyte membrane thus have lower cell power density(maximum 200-800 mW/cm^2), direct liquid NH_3 fueled cells may provide power density quite close to the H_2 fueled cell[52-56], that means higher than the expected 75% power density of H_2 cells. While for the cells with very thin electrolyte the cell performance could very close to the theoretical ratio, 0.75:1.0 for a cell with NH_3 and H_2 as fuel, respectively [57].

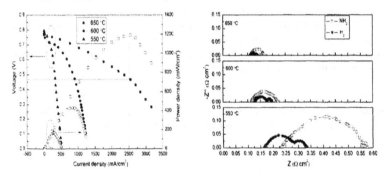

Fig.11 (a) Cell performance with NH_3 fuel at various temperatures, and (b) Impedance spectra of the cell with H_2 and NH_3 under open-circuit conditions.

Shown in Fig. 11 is the result of a cell with SDC electrolyte (10μm thick). We may see that the maximum power density is 1190 mW/cm^2 at 650°C, which is only 63.8% of the value for H_2 fueled cell (see Fig.4) and an obvious concentration polarization behavior is observed. At lower operation temperatures (600–550°C), the V–I curves of the cell are rather strange in that they fall down rapidly at quite small current densities resulting in the peak power densities much lower than the expected. This phenomenon may be attributed to the incomplete decomposition of NH_3 gas in the anode compartment as well as the mass transfer behavior much different from the H_2 cell, because that the cell resistances measured in situ by impedance spectroscopy did not have much difference, as presented in Fig.11(B) [57].

Compared with CMFC-O, CMFCs with doped BaCeO₃ proton electrolytes showed some unique characteristics:

- The cell OCV values are around 1.0V in temperature range of 500- 750°C , which are quite close to theoretical EMF value, indicating less electronic conduction of these proton conductors than SDC or GDC.
- There is almost no activation polarization on V-I curves, implying the quick charge transfer on the electrode interfaces
- The conductivity activation energies obtained from the slopes of V-I curves were similar to SDC cells and even lower in lower temperature range

These properties should be certainly related to the structure and the carrier transfer nature in the CMFC-H systems that are worth to study further [56].

6. Conclusion Remarks

The article has recalled the research activities and the progresses on SOFC/CMFCs at USTC and the following remarks are made:

(1) Solid electrolyte materials as the core material of SOFCs have been extensively studied, including salt-oxide composites, doped ceria and doped BaCeO₃ proton conductors. Among them proton conductive materials including chloride –ceria composite and doped barium cerates have some excellent characteristics worth to investigate further. For commercialization of SOFCs/CMFCs, YSZ, SDC as well as the newly developed doped BaCeO₃ are preferred.

(2) CMFCs with thin membrane electrolyte (YSZ, SDC or doped BaCeO₃) in thickness of around 10μm have been routinely fabricated by cost effective ceramic processing and showed fairly good performance in intermediate temperature range(600⁻850°C)

(3) The interesting results based on new electrode materials, unique electrode structure designs and nano-technique processing have extensively improved the cell performance and studies are on going.

(4) An advanced tubular CMFC design was proposed, and the cost effective fabrication process as well as the interconnect ceramics with high performance have been developed in order to promote the CMFC commercialization.

(5) NH₃ fueled SOFCs/CMFCs with ether oxide ion or proton electrolytes have demonstrated satisfied performance and would be great of promise to perform the distributed CMFC devices without the need to concern the problems such like coking on Ni-based anode and the degradation for doped Barium cerate based CMFCs.

ACKNOWLEDGEMENTS

The authors wish to acknowledge the funding for this work from National Natural Science Foundation of China (50572099, 50602043 and 50730002) and the National High Technology Research and Development Program of China (2007AA05Z157 and 2007AA 05Z151).

REFERENCES

[1] Guangyao Meng, Wanyu Liu, Dingkun Peng, Ionics 4(5-6)451-463, 1998

[2] Shanwen Tao and Guangyao Meng, J. Materials Science Letters 18(1999)81-84

[3]S. W. Tao, Z.L. Zhan, P. Wang, G.. Y. Meng, Solis State Ionics 116(1999)29-33

[4] Q.X.Fu, S.W.Zha, D.K.Peng, G.Y.Meng, B.Zhu, J. Power Sources 104 (2002)73-78

[5] Q.X.Fu, D.K.Peng, G.Y.Meng, B.Zhu, Materials letters 53(2002)186-192

[6] G.Y. Meng, Q/X/ Fu, S.W.Zha et al, Solid State Ionics 148(2002)533-537

[7] R.Doshi, V.L. Ricjhards at al, J.Elctrochemi. Soc. 146(1999)1273.

[8] Q. X. Fu, C. R. Xia, S. W. Zha, J. G. Cheng, R. R. Peng, D. K. Peng and G. Y. Meng, Key Engineering Materials, 224/226 (2002)159

[9]S.W Zha, J.G. Cheng, G.Y. Meng, Material Chemistry and Physics, 77 (2003) 594-597

[10]S.W Zha, C.R. Xia, X.H. Fang et al , Ceramic International, 27 (6): 649-654 2001

[11] Q.X. Fu, X.Y. Xu, D.K. Peng, at al. Journal of Materials Science, 38(2003)2901-2906

[12] S.W. Zha, C.R. Xia, G.Y. Meng, J. Appl. Electrochem, 31 (2001) 93-98

[13] S.W Zha, C.R. Xia, G.Y. Meng,

[14] S.W. Zha, Q.X. Fu; C.R. Xia, G.Y. Meng, Material Letters 47 (6) (2001) 351-355

[15] S.R. Gao, R.Q. Yan, W. Lai, Z. Lang, X.Q. Liu, G.Y. Meng,Chemical Journal on Internet, 4 6 2002 16030

[16]J.G. Cheng, S.W. Zha, J. Huang, X.Q. Liu, G.Y. Meng, Materials Chemistry and Physics 78(2003)791-795

[17]X.H. Fang Q.Dong Y.H. Shen D.K.Peng et al. J. USTC 32 (2002)332-337

[18] R.R. Peng, C.R. Xia, X.H. Fang, G.Y. Meng, D.K. Peng, Materials Letters, 56 (6)(2002): 1043-1047

[19] J.J. Ma, C. R. Jiang, G. Y. Meng, X. Q. Liu, J. of Power Sources, 162(2006) 1082-1087

[20] C.R Jiang, J.J Ma, X.Q. Liu, G.Y. Meng, J. Power Sources 165(2007)134-137

[21] H.B. Li, C.R. Xia, X.H. Fang, X. He, G.Y. Meng, Key Engineering Materials Vols.280-283(2005) pp.779-784

[22] R.R. Peng, C.R.Xia, X.Q.Liu, D.K. Peng, G.Y. Meng, Solid State Ionics, 152-153(2002)561-565

[23] X.Y. Xu, C.R. Xia, D.K. Peng, Ceramic International 31(2005)1061-1064

[24]Y.L. Zhang, J.F. Gao, G.Y. Meng, X.Q. Liu,J. of Applied Electrochemistry, 34(2004)637-641

[25] W.T. Bao, Q.B. Chang, G.Y. Meng, Journal Membrane Science 259(2005)103-105

[26] R.Q. Yan, D. Ding, B. L., M. F. Liu, G.Y. Meng, X.Q. Liu, J. Power Sources 164 (2007) 567–571.

[27] M.F. Liu, D. H. Dong, X.Q. Liu, G.Y. Meng, J. Power Sources, under review 2008

[28]J.F. Gao, X.Q. Liu, D. Peng, G.Y. Meng, Catalysis today 82 (2003)207

[29]Y.H. Chen, J.Y. Wei, X.Q. Liu, G.Y. Meng, Functional Mater35 6 2005 865

[30]Y.H. Chen, J.Y. Wei, X.Q. Liu, G.Y. Meng, J. Inorg. Chem.21 5 2005 673

[31] J.Gui Cheng, H.B. Li, X.Q. Liu, G.Y. Meng, J. USTC, 32(2002)253-258

[31] X.H. Fang, G.Y. Zhu, C.R. Xia, G.Y. Meng, Solid State Ionics 168(2004) 31

[32]Y.H. Yin, W. Zhu, C.R. Xia, G.Y. Meng, J. Power Sources, 132(1-2)36-41(2004)

[33] Y.H Yin, W. Zhu, C.R Xia, C. Gao, G.Y,Meng, J. Appl. Electrochem. 34 (12): (2004)1287

[34]W Zhu, CR Xia, J Fan, R.R Peng,G.Y Meng.J Power Sources, 160 (2006) 897

[35]F. Zhao, R.R. Peng, C.R. Xia, Materials Research Bulletin, 43(2008)370

[36]G.Y. Meng, J.F. Gao, D.K. Peng.X.Q. Liu, China patent: ZL02113198.8

[37]D.H. Dong, J.F. Gao, X.Q. Liu, G.Y. Meng, J. of Power Sources, 165(2007)217-223

[38] D.H. Dong, M.F. Liu, K. Xie, X.Q. Liu, G.Y. Meng J of Power Sources, 175(2008) 201

[39] D.H. Dong, M.F. Liu, K. Xie, J.F. Gao, G.Y. Meng J. of Power Sources, 175(2008) 272

[40]D.H. Dong, J.F. Gao, M.F. Liu,.G.B.Chu, J. Diwu, X.Q. Liu, G.Y. Meng,
Journal of Power Sources, 175(2008) 436-440

[41]M.F. Liu, D.H. Dong, L. Chen. X.Q. Liu, G.Y. Meng, J. of Power Sources, 176(2008) 107

[42]H.H Zhong, X.L. Zhou, X.Q. Liu, G.Y. Meng, Solid State Ionics, 176(2005)1057-1061

[43] X.L Zhou, J.J. Ma, F.J. Deng, G.Y. Meng, X.Q. Liu, J. of Power Sources, 162(2006) 279

[44] X.L Zhou, F.J. Deng, M.X. Zhou, G.Y. Meng, X.Q. Liu, J. of Power Sources, 164(2007) 293 [45] X.L Zhou, F.J. Deng, M.X. Zhou, G.Y. Meng, X.Q. Liu, Mater. Res. Bull., 42(2007)1582

[46] X.L Zhou, M.X. Zhou, F.J. Deng, G.Y. Meng, X.Q. Liu, Acta Materialia,55(2007)2113-2118 [47] X.L Zhou, J.J. Ma, F.J. Deng, G.Y. Meng, X.Q. Liu, Solid State Ionics 177(2007)3461-3466

[48] H.H Zhong, X.L. Zhou, X.Q. Liu, G.Y. Meng, Journal of Rare Earths, 23(1)(2005) 36

[49]M.F. Liu, L. Zhao, D.H. Dong, S.L Wang, J. Diwu, X.Q. Liu, G.Y. Meng,Journal of Power Sources, In Press, Corrected Proof, Available online 23 November 2007

[50]Songlin Wang, Mingfei Liu, Yingchao Dong, Kui xie, Xingqin Liu and Guangyao Meng,Materials Research Bulletin, In Press, Accepted Manuscript, Available online 7 Nov. 2007

[51] Shanwen Tao and John T. S. Irvine, J. Electrom. Soc. 151(2) (2004)A252-259.

[52]Q.L. Ma, R.R. Peng, L. Z. Tian, G.Y. Meng Electrochem. Comm., 8(11)(2006) 1791

[53]Q.L. Ma, R.R. Peng, Y.J. Lia, J.F. Gao, G.Y. Meng, J. Power Sources, 161(1)(2006) 95

[54]Q.L. Ma, J.J. Ma ,S. Zhou, J.F. GAO, G.Y. Meng, J. Power Sources, 164(2007) 86

[55] G.Y. Meng , G.L. Ma, Q.L. Ma, R.R.Peng ,X.Q. Liu, Solid State Ionics 178(2007)697

[56]K. Xie, Q.L. Ma, B. Lin, Y.Z. Jiang, J.F. Gao, X.Q. Liu, G.Y. Meng,Journal of Power Sources, 170(2007)38-41

[57]G.Y. Meng, C.R Jiang, J.J. Ma, Q.L Ma , X.Q. Liu, J. Power Sources, 173(2007) 189

[58]A. Wojcika, H. Middletona, I. Damopoulosa, J. Van herle, J. Power Sources, 118(2003)342.

Cell and Stack Development and Performance

DEVELOPMENT OF MICRO TUBULAR SOFCS AND STACKS FOR LOW TEMPERATURE OPERATION UNDER 550°C

Toshio Suzuki, Toshiaki Yamaguchi, Yoshinobu Fujishiro, and Masanobu Awano
National Institute of Advanced Industrial Science and Technology (AIST)
2266-98 Anagahora, Shimo-shidami, Moriyama-ku,
Nagoya, 463-8560, Japan

Yoshihiro Funahashi
Fine Ceramics Research Association (FCRA)
2266-99 Anagahora, Shimo-shidami, Moriyama-ku,
Nagoya, 463-8561, Japan

ABSTRACT

Micro tubular SOFCs have successfully demonstrated their advantages over conventional (planar) SOFCs. The advantages are high thermal stability during rapid heat cycling and high electrode area per volume, which enable micro SOFC systems applicable to portable devices and auxiliary power units for automobile. In this study, fabrication and integration of micro tubular SOFCs under 1 mm diameter was discussed. So far, the cell performance of a 1 cm long tubular single cell with diameter 0.8 mm, where the cathode covers 5 mm of the tube, was shown to be over 0.12 W at 550 °C operating temperature. To bundle the micro tubular cells, a cube shaped SOFC bundle design was considered and fabricated, which consists of micro tubular SOFCs and cathode matrices with high porosity and reasonable electrical conductivity. In addition, a calculation model of the cube bundle to estimate current collecting loss was discussed.

INTRODUCTION

Recently, numbers of great works have been shown especially for lowering operating temperature of SOFC under 600 °C. One of examples is reported by J. Yan, $et\ al.$ showing that the advantages of using $La_{0.9}Sr_{0.1}Ga_{0.8}Mg_{0.2}O_{3-\delta}/Ce_{0.8}Sm_{0.2}O_{2-\delta}$ composite electrolyte and high cell performance of over 600mW/cm^2 at 500 °C [1]. Z. Shao $at\ el.$ achieved over 400 mW/cm^2 of cell performance at 500 °C by introducing $Ba_{0.5}Sr_{0.5}Co_{0.8}Fe_{0.25}O_{3-\delta}$ as a cathode material, even though a typical electrolyte material, $Ce_{0.8}Sm_{0.2}O_{2-\delta}$, was used for an electrolyte [2].

Despite the progress in material development, there are still drawbacks such as slow start up time and low robustness under rapid heat cycles. A way to avoid the drawbacks is to change the cell shape into the tubular design. Tubular SOFCs have successfully demonstrated their advantages over conventional (planar) SOFCs in regards to thermal stability during rapid heat cycling and possible high electrode area per volume, which enable SOFC systems applicable to portable devices and auxiliary power units for automobile.

The history of tubular design in SOFC is relatively old. Actually, in 1980 Siemens Power Generation demonstrated the tubular SOFC principle and successfully conducted long-term operation over 70000 h. Then, small-scale tubular SOFCs with 2 mm diameter was shown by Kendall and Palin [3], and Yashiro $et\ al.$ [4] showed that they could endure thermal stress caused by rapid heating up to operating temperature.

Our study aims to develop a fabrication technology for micro tubular SOFCs and their stacks. Previously. we proposed a new deign of micro tubular cell stacks using porous cathode matrices as a support of the cells for assembling a cube- type SOFC bundle [5]. Figure 1 show the schematic image of new stack design using the cube-type SOFC bundles with current collectors. In this study. fabrication and an integration method for micro tubular SOFCs under 1 mm diameter was discussed and the cube bundle design was considered and fabricated. In addition. the current collecting loss due to cell configuration in the cathode matrices was discussed using a calculation model to design high efficient cube-type SOFC bundles with micro tubular SOFCs.

EXPERIMENTAL
Fabrication of the micro tubular SOFCs
Anode supported tubes were prepared from NiO powder (Seimi Chemical co., ltd.), $Gd_{0.2}Ce_{0.8}O_{2-x}$ (GDC) (Shin-Etsu Chemical co., ltd.). a pore former (poly methyl methacrylate beads (PMMA). Sekisui Plastics co.. ltd.). and a binder (cellulose, Yuken Kogyo co., ltd.). These powders were mixed for 1 h by a mixer 5DMV-rr (Dalton co.. ltd.). After adding a proper amount of water, the mass was stirred for 30 min in a vacuumed chamber, and then, was left for aging for over 15 h. The mass was extruded into anode tubes from a metal mold using a piston cylinder type extruder (Ishikawa-Toki Tekko-sho co.. ltd.).

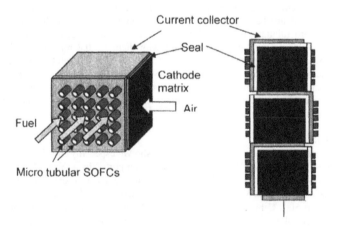

Fig. 1 Schematic image of the micro SOFC cube bundle and stacks

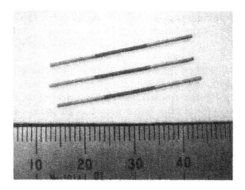

Fig. 2 Image of the 0.8 mm diameter micro tubular SOFCs prepared in this study.

A slurry for dip-coating the electrolyte was prepared by mixing the GDC powder, solvents (methyl ethyl ketone and ethanol), a binder (poly vinyl butyral), a dispersant (polymer of an amine system) and a plasticizer (dioctyl phthalate) for 24 h. The anode tubes were dipped in the slurry and coated at the pulling rate of 1.5 mm/sec. The coated films were dried in air, and sintered at 1400°C for 1 h in air. The anode tubes with electrolyte were, again, dip-coated using a cathode slurry, which was prepared in the same manner using $La_{0.6}Sr_{0.4}Co_{0.2}Fe_{0.8}O_{3-y}$ (LSCF) powder (Seimi Chemical, co., ltd.), the GDC powder, and organic ingredients. After dip-coating, the tubes were dried and sintered at 1000°C for 1 h in air to complete a cell. Figure 2 shows an image of complete 0.8 mm diameter cells.

The cell performance was investigated by using a potentiostat (Solartron 1296). The cell size was 0.8 mm in diameter and 10 mm in length with cathode length of 5 mm, and the effective cell area was 0.126 cm^2. Ag wires fixed with Ag paste were used for collecting current from anode and cathode sides, which were both fixed by an Ag paste. The current collection from the anode side was made from an edge of the anode tube. Hydrogen (humidified by bubbling water at room temperature) was flown inside of the tubular cell at the rate of 5 mL min^{-1}. The cathode side was open to the air without flowing gas.

Fabrication of the micro tubular SOFC bundle

The cathode matrices were made from an LSCF powder (Daiichi Kigenso Kagaku Kogyo co., ltd.), a binder (cellulose) and a pore former (70vol.% of PMMA). They were mixed using the same manners as for the anode support preparation. In order to control the microstructure of the cathode matrix, LSCF starting powders with 0.05, 2, and 20 μm particle sizes were tested. The matrices were prepared by extruding the clay from a metal mold using a screw cylinder type extruder (Miyazaki Tekko co., ltd.). The extruded matrices were cut into 1 cm long pieces, dried and sintered at 1400 °C in air. The shape of the cathode matrix was 1 cm wide, 1 cm long and 2 mm thick, with 5 grooves of 1 mm width on the surface, cf. Figure 3a.

The gas permeability test of the cathode matrices was carried out using the experimental apparatus referred to in the literature [5]. The cathode matrices without the grooves were used for this measurement. Each sample was a plate whose size is 2 cm in width, 2 cm in length and 2.5

mm in thickness, and the area of gas permeability was 0.79 cm^2. The electric conductivity of the cathode matrices was investigated by using a DC power supply (ADVANTEST R6234) and Digital multi meter (KEITHLEY 2700) in DC 4 point probe measurement.

Figure 3 (a) shows the procedure of assembling the cube bundle. A bonding paste for assembling the cube bundle was prepared by mixing the LSCF powder (Seimi chemical co., ltd.), a binder (cellulose), a dispersant (polymer of an amine system), and a solvent (diethylene glycol monobutyl ether). The paste was screen-printed on the surface of the cathode matrices, followed by the placement of the tubular cells before the paste was dried. After the cube bundle was assembled by sandwiching the cathode matrices and the tubular cells, the cube bundle was sintered at 1000 °C for 1 h in air. The complete cube bundle with 25 tubular cells are shown in Fig. 3 (b).

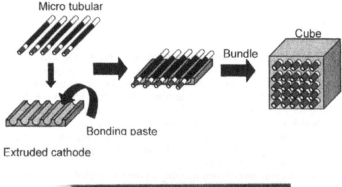

Fig. 3 (a) Fabrication procedure of the cube bundle and (b) an image of the cube bundle with 25-8 mm diam. micro tubular SOFCs.

RESULTS AND DISCUSSION
Performance of the tubular cell

Figure 4 shows the result of fuel cell tests for the micro tubular SOFC. As can be seen, the peak powers of 36, 81 and 131 mW were obtained at 450, 500, and 550 °C operating temperature, respectively. Note that the open circuit voltages dropped from 0.89 to 0.83 V as the operating temperature increased from 450 to 550 °C. This can be related to an increased electronic conductivity in the ceria electrolyte due to reduction and/or physical gas leakage through gas sealant, which need to be improved. There are several efforts on improving OCV for ceria based electrolyte SOFCs by introducing a second layer electrolyte [6, 7]. Such techniques can also be applied to improve the OCV of the micro tubular cells.

The performance of the cell per weight turned out to be 4.55, 10.5 and 16.9 W/g at 450, 500, and 550 °C, respectively. Since the cell components were prepared from conventional materials [8], these outstanding performances seem to be a result of an improved anode microstructure for gas diffusion and electrochemical reactions.

Using the micro tubular SOFCs, the cube bundle (1 cm^3) shown in Fig. 3 (b) is expected to have power output of 1.8, 4.0 and 6.5 W at 450, 500, and 550 °C, respectively.

Characterization of the cathode matrices

Table 1 shows the characteristics of the cathode matrices; the porosity, the gas permeability and the electrical conductivity. Each sample is named after the grain size of the starting LSCF powder. The cathode matrices prepared in this study has open porosity of 60~80% and electrical conductivity of 50~140 S/cm. As can be seen, the initial grain size of the starting LSCF powder strongly affected the properties, especially the gas permeability increased drastically by using lager grain sized LSCF powders. On the other hands, use of large LSCF powder sacrificed the electrical conductivity which will increase of the current collecting loss. These results will be used for further discussion, later.

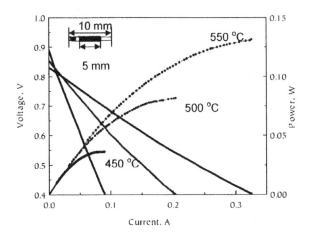

Fig. 4 Results of the cell performance test for the single tubular SOFC (0.8 mm diam.)

Table I: Characterization of cathode matrices

	LSCF0.05	LSCF2	LSCF20
Grain size of LSCF (μm)	0.05	2	20
Porosity (%)	60.7	68.6	78.2
Gas permeability (ml cm cm^{-2} sec^{-1} Pa^{-1})	8.5 X 10^{-6}	8.4 X 10^{-5}	6.2 X 10^{-4}
Conductivity (S/cm)@500°C	139	108	49

Amount of PMMA 70vol.%   Sintering temperature 1400°C

Estimation of the cube bundle current collecting loss

Since all micro tubular SOFCs in the cube bundle were connected in parallel, which leads to low total cell resistance, and the cathode matrix has relatively high electrical resistance, the current collection from the cube bundle would become a crucial issue. Thus, it is important to understand the effect of current collection on the performance of the cube bundle. For this purpose, a current collecting model was introduced to calculate the performance loss due to the current collection from the cube bundle.

Figure 5 (a) shows the schematic image of the cube bundle; a current collector was placed at a side of the cube bundle. The current flow in the tubular cells was assumed to be constant and uniform, as well as the current flow in the cube in the y direction. Thus, we extracted a 1 x 5 tubular cell layer from the cube bundle for further calculation as shown in Fig. 5(a).

The cathode matrix with the anode tubes was sliced as shown in Fig. 5(b), and each of them was assigned to an equivalent circuit (highlighted in Fig.5 (b)). Especially, each anode tube part was divided into N (N=2000) elements to minimize the error of this calculation [9]. Δx, ΔR_{cell}, $\Delta R_{cathode}$ shown in Fig.5 (b) are respectively the width of sliced cathode matrix (Δx = tube diameter /N), the resistance of the tubular cell and the cathode matrix in the slice, respectively. Values of these resistances were estimated from the peak power density of the tubular cell shown in Fig. 4 (at 500 °C) and the conductivity shown on Table I (LSCF0.05, LSCF2, LSCF20), respectively. In this calculation, the current collecting resistance of the anode tube was not included and separately discussed elsewhere [9]. Current collecting loss was determined using the following relation.

$$\text{Current collecting loss} = (R_N - R_{total\ cell})/ R_N \qquad (1)$$

R_N is the resistance of total cube bundle including the resistances of the cathode matrix and the tubular cells, and $R_{total\ cell}$ is the resistance of the tubular cell (5 cells in parallel), respectively.

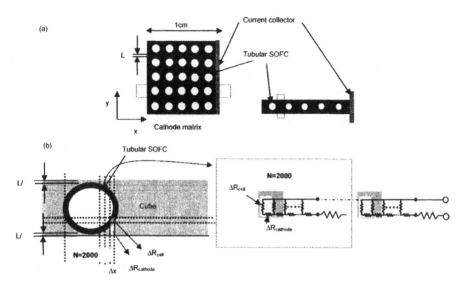

Fig. 5 (a) A calculation model and (b) an equivalent circuit for estimating current collecting loss.

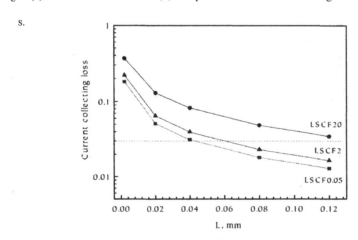

Fig. 6 Current collecting loss as a function of the distance between tubular cells in the cube bundle at 500 °C for different cathode matrices.

Figure 6 shows the current collecting loss as a function of distance between tubes as shown in Fig. 5 (a) at 500 °C. As can be seen, the loss dirctly influenced by the distance between tubes. The current collecting loss estimated for the fabricated cube bundle is less than 3 % when LSCF0.05 and LSCF2 were used for the cube bundles and the distance L > 0.06 mm. The results indicated that further accumulation of tubular cells in the cathode matrices (reduce L) to obtain higher power density in volume will lead to increase the current collecting loss. There are several possible ways to improve the cube bundle performance for bundles containing more than 25 cells, for expamle, use of alternative materials (higher conducting materials) for cathode matrices. In any case, carefull consideration is needed for further improvement of the cube bundle.

CONCLUSION

High performance, sub-millimeter diameter micro tubular SOFCs have been successfully fabricated. A new SOFC bundle design using the micro tubular cells and cathode matrices was also proposed and fabricated. The cathode matrices prepared in this study has open porosity of 60~80% and electrical conductivity of 50~140 S/cm. The current collecting loss of the cube bundle was calculated as a function of a distance between micro tubular SOFCs in the cube bundle. The results showed that the present model (5 x 5 tubular cells) had less than 3 % of performance loss by selecting appropriate materials for cathode matrices. However, the performance of the cube bundle is dependent on the size of the bundle, and further consideration are needed for upscaling.

ACKNOWLEDGEMENT

This work had been supported by NEDO, as part of the Advanced Ceramic Reactor Project.

REFERENCES

[1] J. Yan, H. Matsumoto, M. Enoki and T. Ishihara. *Electrochem. Solid-Sate Lett.,* 8 (8), **2005,** A389-A391.
[2] Z. Shao and S. M. Haile. *Nature.* 431 1, **2004,** 70-173.
[3] K. Kendall and M. Palin. *J. Power Sources,* 71, **1998,** 268-270.
[4] K. Yashiro, N. Yamada, T. Kawada, J. Hong, A. Kaimai, Y. Nigara and J. Mizusaki. *Electrochemistry.* 70, No.12, **2002,** 958-960.
[5] Y. Funahashi, Y. Shimamori, T. Suzuki, Y. Fujishiro and M. Awano. *J. Power Sources,* 163, **2007,** 731-736.
[6] A. Tomita, S. Teranishi, M. Nagao, T. Hibino and M. Sano. *J. Electrochem. Soc.,* 153, **2006,** A956-A960.
[7]. E.D. Wachsman, P. Jayaweera, N. Jiang, D.M. Lowe and B.G. Pound. *J. Electrochem. Soc.,* 144. **1997,** 233-236.
[8] P. Bance, N. P. Brandonm B. Girvan, P. Holbeche, S. O'Dea and B. C. H. Steele. *J. Power Sources,* 131, **2004,** 86-90.
[9] T. Suzuki, T. Yamaguchi, Y. Fujishiro and M. Awano. *J. Power Sources,* 163, **2007,** 737-742.

THE PROPERTIES AND PERFORMANCE OF MICRO-TUBULAR (LESS THAN 1MM OD) ANODE SUPPORTED SOLID OXIDE FUEL CELLS

N. Sammes[1,2], J.Pusz[2], A. Smirnova[2], A. Mohammadi[2], F. Serincan[2], Z. Xiaoyu[2], M. Awano[3], T. Suzuki[3], T. Yamaguchi[3], Y. Fujishiro[3], Y. Funahashi[4]

[1]The Department of Metallurgical and Materials Engineering, Colorado school of Mines, Golden, CO, USA
[2]Connecticut Global Fuel Cell Center, University of Connecticut, Storrs, CT 06269, USA
[3]Institute of Advanced Industrial Science and Technology (AIST), Nagoya, Japan;
[4]Fine Ceramics Research Association, Nagoya, Japan.

ABSTRACT

The sub-millimeter range micro-tubular SOFCs have been modeled, designed and tested in the intermediate temperature range of 450~550°C. The results of a two-dimensional, steady state, non-isothermal computational model incorporating radiation, heat transfer, flow and transport of species in the porous gas diffusion electrodes are presented and compared to the experimental data. Considering different current collecting techniques the model estimates the loss of performance for a single cell as a function of anode tube length and thickness. The results showed that for the 0.8 mm diameter tubular SOFC with a length of 1.2 cm the performance loss is 0.8, 2.0, and 4.6 % at 450, 500, and 550°C respectively. The mechanical tests with electrolyte coated Ni-GDC anode pellets indicate that the micro- and nano-hardness of GDC is comparable or higher than that of YSZ. In the intermediate temperature range of 450~550°C the performance of 0.8 mm diameter micro-tubular ceria-based SOFCs was found to be between 110~350 mWcm^{-2}.

INTRODUCTION

Solid Oxide Fuel Cells (SOFCs) consisting of durable ceramic components can be considered nowadays as the most promising fuel cell systems for alternative source of energy due to their high electrical conversion efficiency, superior environmental performance, and fuel flexibility[1-3]. In comparison to conventional SOFCs, micro tubular SOFCs[4-6] have many advantages such as high resistance to thermal shock and higher power densities reaching 0.3Wcm^{-1} at 550°C[7]. Recent improvements of the mechanical and electrochemical properties at reduced operating temperatures allows the use of cost-effective materials for interconnects and balance of plant[8-9].

Traditionally, Yttria stabilized Zirconia (YSZ) and NiO-YSZ cermets have been used as electrolyte and anode materials in SOFC systems, respectively. Electrochemical and mechanical properties of these materials under different operational conditions have been extensively studied[10-11] and optimized by changing the cermet structure and particle size distribution[12-14]. To reduce the operating temperature, and increase cell performance, alternative materials for the cell components have been developed, such as perovskite single phase and composite anode and cathode materials and electrolytes operating in the intermediate temperature range[15-17].

The present paper is focused on modeling, mechanical, and electrochemical properties of the micro-tubular SOFCs and SOFC cube stacks. To predict a microtubular SOFC performance, a two-dimensional (2D) model was considered followed by the results of mechanical and electrochemical testing. The attention was paid to the development of materials for electrodes and electrolytes that lower the operating temperatures to below 650°C, as well as to the development of manufacturing processes for sub-millimeter tubular SOFC cells for arrangement and integration at the micro level.

The results of a micro-tubular SOFC and a cube-type cell stack module fabricated and tested to check its applicability for commercialization of an auxiliary power unit (APU) and a stationary small power source system are presented. The micro-tubular SOFCs with high energy efficiency, quick start-up and shut-down performance and low production cost are described in addition to the mechanical properties and fabrication technology data for the SOFC materials operating in the temperature range of 450-650°C.

MODELING AND SIMULATION

The modeled geometry is based upon micro tubular cells fabricated as a part of the Advanced Ceramic Reactor Project (ACRP) initiated by New Energy and Industrial Technology Development Organization (NEDO) of Japan. The single cell tests are assumed to be carried out in a firebrick furnace in which the heat is supplied by the wires surrounding the cell so that there is an even peripheral heat distribution. The modeled domain covers only a portion of the test geometry for a couple of reasons. Firstly, due to the high aspect ratio of the cell length to the electrolyte thickness, the modeled length is taken as quarter of the actual cell length. Secondly, for the sake of simplicity while handling the radiation heat transfer, the model domain does not include the heating elements. Since the test geometry employs axial symmetry, contiguous rectangles which revolve around an axis are used to represent the cell.

While defining the geometry, parameters (Table 1) are determined either by direct measurements or from the SEM images provided in the work of Suzuki et al.[18].

Table I. Geometry Parameters

Description	Value	Reference
Interior radius of the anode tube, r_i	1.0 mm	direct measurement
Exterior radius of the anode tube, r_0	1.3 mm	direct measurement
Electrolyte Thickness, t_e	25 μm	[18]
Cathode thickness, t_c	50 μm	[18]
Cathode current collector thickness, t_{cc}	50 μm	[18]
Length of the actual cell, l	1.0 cm	assumed
Air channel thickness, t_{ac}	2.0 mm	assumed

The presented model takes into account the steady state heat, mass, momentum and charge transfer phenomena. Heat equation is used with both convection and conduction mode in all the domains except the electrolyte for which only the conduction mode heat transfer is modeled. For the anode, cathode and current collector regions, an effective heat conduction coefficient is suggested to combine the heat transfer in solid and fluid phases in a single equation. Fick's law is used to govern mass transfer. Momentum transfer is governed by the Navier-Stokes equation in the channels, whereas Darcy's law is used for porous regions such as the anode, cathode and current collectors. Finally, electronic and ionic charge transfers are modeled by Ohm's law.

The heat, mass, momentum, and charge transfer governing equations are solved using a commercial computational fluid dynamics package COMSOL, which is equipped with predefined partial differential equations. Along with a direct UMFPACK linear system solver, nonlinear stationary solvers are used[19]. With 7878 triangular mesh elements, the solution becomes satisfactory in the error tolerance band of 10 considering 70126 degrees of freedom in the domain. Weak formulation is used for solving the model, which generates the exact Jacobian for the fast convergence of nonlinear models.

The model presented in the preceding sections is validated by the experimental data of Suzuki et al.[18]. The operational parameters and material properties used in the model are taken to be the same as the actual ones for comparison. The ones that are not given or not found in the literature are chosen with reasonable assumptions. In Figure 1a, polarization curves obtained from the model and the experiments are compared. It can be seen that the model results correlate reasonably well with experimental data. The comparison of the power density curves as shown in Figure 1b is also promising. However, the model overestimates the actual results in mid-range current densities.

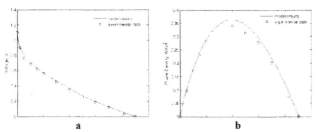

a   b

Figure 1. Comparison of a) polarization curves and b) power characteristics obtained from the model (solid line) and acquired from experiments of Suzuki et al.[18].

It can be seen from the polarization curve that at higher current densities the curve has a concave up characteristic which is not an usual result for fuel cell applications. For an ordinary fuel cell, at higher current densities, concentration losses are dominant which is reflected as a concave down curve. As is discussed in the introduction section, the concentration losses in micro scaled fuel cells would not be as notable as those observed in macro scaled ones due to significantly smaller concentration gradients. In addition, at higher current densities, the temperature of the fuel cell increases because of ohmic resistance. The increase in local temperature (which is the temperature used throughout this work) improves the cell performance due to the fact that conductivity of the electrolyte increases with increasing temperature. In Figure 2a the temperature distribution inside the cell for different operating voltages, marked on the vertical axis, can be seen. Corresponding ionic conductivities of the gadolinia-doped ceria (GDC) electrolyte can be found in Figure 2b. Ionic conductivity of GDC as a function of temperature was obtained from[20].

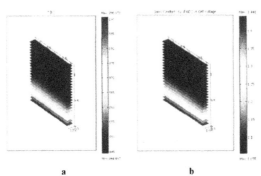

a   b

Figure 2. a) Temperature distribution inside the cell for different operating voltages indicated in vertical y-axes and b) Corresponding values of the GDC electrolyte ionic conductivity.

The temperature distribution inside the modeled domain is shown in Figure 3a. The effect of the radiation is seen as the temperature profiles evolve from the regions closer to the most right boundary. As discussed in the previous sections some portion of the energy irradiated by the wires is absorbed by the cell while the rest is given to the fluid flow.

a   b

Figure 3. a) Temperature distribution inside the computational domain at the operating voltage of 0.7 V (min 473 K, max 1053 K), contours show the temperature profiles and b) Temperature distribution inside the cell at the same operating point (min 850.63 K, max 853.48 K)

This consequence is reflected on the temperature distribution as the temperature profiles have a shape of a bell pointing to the outlet of the channel, which explains that fluid regions closer to the heating elements acquire energy by radiation, whereas the regions closer to the cell acquire energy by convection. Since these two distinct heat transfer mechanisms transmit different amounts of energy to the fluid, the bell shaped profile is inclined towards the cell meaning flow gains more energy from radiation. Hereby, it could be implied that preheating of the gases before entering the furnace would result in an increase in cell performance because in the case of preheated gases, less energy would be released from the cell to the flow.

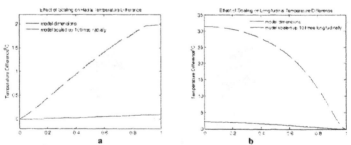

a   b

Figure 4. a) Temperature difference in the radial direction when the dimensions are scaled up ten times and b) Temperature difference in the longitudinal direction when the dimensions are scaled up ten times.

Corresponding temperature distribution inside the cell is plotted in a different figure (Figure 3b) to distinguish the small temperature difference compared to the overall distribution. As discussed in the previous sections. the temperature difference inside the cell is significantly small for a micro scaled fuel cell. Only less than 3°C of a difference is observed. For a more meaningful measurement, the longitudinal temperature difference for a unit length is 12°C/cm and radial temperature difference for unit length is 3°C/cm. These values are consistent with the order of magnitude reported in the previous work of our group[21].

For the quantitative evaluation of the micro-tubular SOFC. the model geometry was scaled up 10 times in the radial direction, and 10 times in the longitudinal direction. With the same amount of energy flux given to the system and the same flow rates. the results are plotted (Figure 4) in terms of temperature differences as functions of normalized cell thickness and length. In both cases, temperature differences increase notably when cell geometry is enlarged. Enlarging the geometry 10 times resulted in a temperature difference of 20.44 times more than that found for the original dimensions, whereas longitudinal scaling resulted in a temperature difference of 14.75 times higher.

In the context of miniaturization it should be stated that for a micro scaled fuel cell the concentration gradients are small enough to decrease mass transport losses notably. Figure 5 shows the hydrogen and oxygen molar concentrations for the operating voltages of 0.8. 0.45 and 0.1 V. Comparing 0.8 V and 0.1 V it can be deduced that there is 39 % decrease in H_2 concentration whereas the decrease is only 9% for O_2. This result is noteworthy because it is known that for macro-scaled fuel cells either, SOFC or PEMFC. at higher current densities severe mass transport limitations are predominant. In some cases, with the diminishing concentration of the reactants, starvation of the cell takes place. The consequence of the small gradients in miniaturized cells is shown in the polarization curve characteristics. as discussed previously.

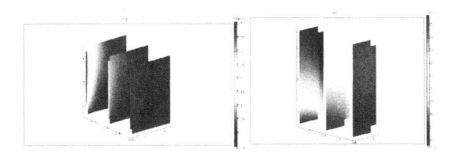

a   b

Figure 5. a) Hydrogen concentration distribution for operating voltages of 0.8, 0.45 and 0.1 V. (min 15.02 mol/m^3, max 25.75 mol/m^3). contours show the temperature profiles; b) Oxygen concentration distribution for the same operating voltages (min 4.952 mol/m^3, max 5.408 mol/m^3).

The effect of scaling on hydrogen concentration can be seen in Figure 6. For the cell operating at 0.7 V, the hydrogen concentration profiles along a vertical line in the anode are compared for the original dimensions and dimensions 10 times scaled in the radial direction. The drop in the concentration along the profile is three times more than for the cell with larger dimensions.

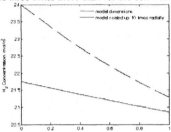

Figure 6. Comparison of the hydrogen concentration profiles in the anode for the original cell (solid) and the scaled up cell (dashed).

MECHANICAL TESTING

The micro-hardness results for different anode and electrolyte materials are shown in Table II. The GDC powder from Anan Kazei in comparison to the United Ceramics Limited YSZ powder has almost the same mechanical properties. However, the mechanical properties of GDC from DKKK with coarser particle size were even better. Since all different powders were fired under the same conditions, this can be related to the particle size of powders and the effect of grain boundaries impurities known to change sintering properties of ceramic materials. The YSZ nano-powder from TOSOH probably activates another sintering mechanism due to the nano-powder properties well known for their lower sintering temperatures. In this case a strong bonding in the grain boundary regions as well as inside of the grain areas improves the hardness of the pellets (Table III).

Table II. Average Micro-hardness (300 gr load) and Coefficient of Variation.

Material	Hardness (VHN)	Coefficient of Variation (%)
GDC (Anan Kazei)	673.8	10.2
GDC (DKKK)	899.3	7.1
YSZ (TOSOH) nano-crystalline powder	1379.2	4.9
YSZ(United Ceramics Limited)	690.5	3.0
NiO-GDC	1358.3	4.6
Screen Printed GDC on the anode substrate	775.7	4.9

Nano-indentation results for different samples (Table 2) confirm that the mechanical properties of GDC powder is in the same range or better than those of the United Ceramics YSZ powder. For the GDC layer screen printed on the anode support, application of the lower load reduces the depth of penetration of the indenter, which minimizes the effect of substrate on the hardness of the coating. As expected, unlike the micro-hardness results, the hardness and modulus of screen printed GDC is lower than those of the GDC pellet (Figure 7). Different GDC and YSZ pellets were tested using ASTM F 349-78 (1996) standard, however, the data collection from delaminated GDC pellets, made from the coarser powder, was a challenge since they broke before any reliable data collection. Chipped edges, due to high contact stress under the loaded balls, also occurred for some specimens. This issue was solved by applying a thin plastic layer underneath the specimen to reduce the local stress at the contact points.

Table III. Average Nano-hardness (5* and 50**g load), Elastic Modulus, and Standard Deviation.

Material	Hardness (GPa)	Modulus (GPa)	Standard Deviation
GDC** (Anan Kazei) pellet	4.8	129.9	4.3
GDC** (DKKK) pellet	7.7	143.4	0.7
YSZ**(United Ceramics Limited) pellet	4.9	131.2	4.5
GDC* (Anan Kazei) screen-printed layer on the anode substrate	1.9	27.2	2.3

The calculated biaxial flexure strength for GDC was $\sigma = 20$ MPa, however, since the pellet was delaminated this number is much less then the true strength value of GDC pellets. The modulus of rupture of the pellets made from the YSZ nano powder was about 114 MPa. These calculations were based on the peak loads of 60.2N and 297.0N for GDC and YSZ pellets, respectively (Figure 8).

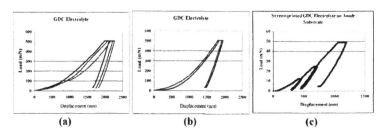

(a)   (b)   (c)

Figure 7. Nano-hardness tests on: (a) Anan Kazei GDC pellets; (b) DKKK GDC pellets; (c) Screen-printed GDC on NiO-GDC anode. The higher displacement in Anan Kazei specimen represents lower hardness of the substrate.

(a)   (b)

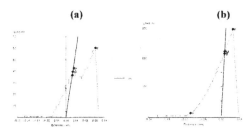

Figure 8. Biaxial Flexure test on GDC (a) and YSZ (b) pellets.

The measured fracture strength of NiO-YSZ powder after different redox cycles are reported in Table IV. It has been shown that there is no significant increase in the fracture strength between the non-reduced and reduced anode. It has also been shown that after the first and third redox cycles the maximum strength resulted in 43% and 77% increase, compared to that of the reduced anode, respectively. Further redox cycles reduced the maximum strength values. However, those values still remained higher than those of the reduced anodes.

Table IV. Average Fracture Strength, Standard Deviation, and Weibull Parameter of the C-ring Anodes.

Cycle	$\sigma_{average}$ (MPa)	Standard Deviation	Weibull Parameter
Non-reduced Anode	65.1	3.8	16
Reduced Anode	69.0	6.7	10
First Redox	98.6	12.5	8
Second Redox	118.0	23.0	5
Third Redox	122.4	5.5	22
Fourth Redox	105.3	9.8	10
Fifth Redox	105.3	15.8	6

The morphology and structure of the anode pellets after indentation, as well as micro- and nano- indentation marks shown in Figure 9, clearly demonstrates the cracks propagated from the micro-indentation tips.

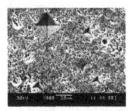 

Figure 9. The SEM micrographs of an anode pellet used for measuring the fracture toughness of the material (a) Micro- and nano-indentation marks and (b) propagated cracks.

SOFC MANUFACTURING AND TESTING

Anode tubes were fabricated using NiO powder (Seimi Chemical co., ltd.). $Gd_{0.2}Ce_{0.8}O_{2-x}$ (GDC) (Shin-Etsu Chemical co., ltd.), poly methyl methacrylate beads (PMMA) (Sekisui Plastics co., ltd.), and cellulose (Yuken Kogyo co., ltd.). After adding the correct percentage amount of water, GDC and NiO powders were mixed using a mixer 5DMV-rr (Dalton co., ltd.) in a vacuumed chamber, producing an extrudate. Tubes were then extruded from the extrudate using a piston cylinder type extruder (Ishikawa-Toki Tekko-sho co., ltd.). An electrolyte was prepared on the surface of the anode tube by dip-coating a slurry consisting of the GDC powder, solvents (methyl ethyl ketone and ethanol), binder (poly vinyl butyral), dispersant (polymer of an amine system) and plasticizer (dioctyl phthalate), and the whole was co-sintered at 1100-1400 °C for 1 h in air at 1400 ^{o}C. The cathode was prepared by dip-coating a slurry of $La_{0.6}Sr_{0.4}Co_{0.2}Fe_{0.8}O_{3-y}$ (LSCF) powder (Seimi Chemical. co., ltd.), the GDC powder, and organic ingredients. After the dip-coating process, the tubes were dried and sintered at 1000°C for 1 h in air. Currently, micro tubular SOFCs with a diameter of 0.8 -2 mm were successfully fabricated using this technique.

The microstructure of the anode tubes with electrolyte sintered at various temperatures was observed by using SEM (JEOL, JSM6330F). The cell performance was investigated by using a potentiostat

(Solartron 1296). The cell size was 0.8 mm in diameter and 8 mm in length with cathode length of 5 mm, whose effective cell area was 0.13 cm^2. Ag wire was used for collecting current from the anode and cathode sides, which were both fixed using Ag paste. Hydrogen (humidified by bubbling water at room temperature) was flowed inside the tubular cell at the rate of 5 mL min^{-1} + Nitrogen 10 $mLmin^{-1}$. The cathode side was open to the air without flowing gas.

Single cell performance test was conducted using wet H_2 fuel in the temperature range of 450~600 ^0C. The power density of the cell calculated from the area of the cathode and was shown in Fig. 11. The performance of the cell per weight turned out to be 4.55, 10.5 and 16.9 W/g at 450, 500, and 550 ^0C, respectively. Since cell components were prepared from typical materials[22], these performances resulted from favorable anode microstructure for gas flow as well as electrochemical reaction of the fuel in the anode. Currently, bundle/stack fabrication technology has been intensively investigated designed for application use, and successful fabrication processes for the cube consisting of 36 cells in 1cm³ was established (Figure 10). In addition, a Honey-comb type micro-tubular SOFCs, which consist of sub-millimeter cells, were demonstrated (Figure 12) and promising results regarding cell performance of over 2W per 1cm³ at 650 and quick start-up (5 min starting from room temperature) were obtained. Further work is now in progress to optimize the cube-like stack.

Figure 10. Micro-tubular SOFC with 0.8 and 1.8 mm diameter, SEM image of the fuel cell cross-section, and a 36 cell stack.

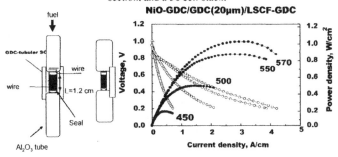

Figure 11. Micro tubular SOFCs configuration and performance characteristics.

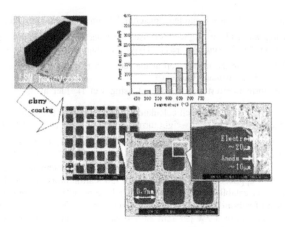

Figure 12. Honeycomb type cathode-supported micro-SOFC integration process and electrochemical properties at different temperatures of operation.

The model cube ($27 cm^3$) with 36 micro tubular SOFCs (2 mm in diameter) was fabricated and is shown in Figure 13. The module demonstrated a high output performance of about 15 W at 650°C and 13.5W below 650°C.

Figure 13. Module assembly and evaluation of prototype module for various applications

CONCLUSION

The modeling, analysis, and testing strategies for SOFCs have been successfully established. The simulation results show interesting phenomena behind the small-diameter SOFC operations that correlates reasonably well with the experimental data. It is observed that at higher current densities there is a slight performance improvement due to ohmic heating which increases the cell temperature and the ionic conductivity of the electrolyte. Significantly small temperature and concentration gradients are observed along both radial and longitudinal directions in the fuel cell due to the micro scales of the system. Comparing the results with those obtained from the model of a ten times larger cell, temperature differences in radial and longitudinal directions in a micro tubular cell found to be around 20 and 15 times less, respectively, than those in a larger cell. This important remark gives rise to the alleviation of the mechanical properties of the cell due to the projected decline in thermal induced stresses.

Mechanical tests performed for determining mechanical properties of NiO-GDC anode and GDC electrolyte materials from different manufacturers indicate that micro- and nano- hardness of GDC is comparable to that of YSZ. However, it should be noticed that the effect of particle size and the firing processes have a large effect on the measured values. Applying the same firing cycle on GDC and YSZ pellets made from powders with the same range particle size shows their hardness as about 900 VHN and 690 VHN, respectively. The cermet NiO-GDC anode also showed good mechanical integrity. Biaxial flexure strengths of these materials were determined, however, there were some surface and bulk issues that should be addressed such as pores or voids, which lead to insufficient internal integrity of the pellets. Further improvement of the testing procedures as well as three- and four-point bend tests can provide more information about mechanical properties with higher precision.

The results of the microtubular SOFC testing indicate that a power density of up to $1 W/cm^2$ can be achieved at 570°C with 20µm GDC electrolyte and composite LSCF/GDC cathode. Honeycomb module demonstrated the values of power density up to $370 W/cm^2$ at 750°C which can be also considered as a significant success toward large-scale SOFC manufacturing.

ACKNOWLEDGMENTS

Financial funding was provided by the New Energy and Industrial Technology Development Organization (NEDO) in the frame of the Advanced Ceramic Reactor Project Contract Number AG060145.

REFERENCES

1. M. C. Williams , J. P. Strakey, W. A. Surdoval, L. C. Wilson, *Solid State Ionics*, **177**, 2039 (2006).
2. M. Masashi, H. Yoshiko, N.M. Sammes, *Solid State Ionics*, **135**, 743 (2000).
3. N.M. Sammes, ed. Fuel Cell Technology: Reaching Towards Commercializtion. Engineering Materials and Processes. 2006, Springer: Berlin.
4. J. Turner, M.C. Williams, K. Rajeshwar, *Electrochem. Soc. Interface*, **13-3**, 24 (2004).
5. K. Kendall and M. Palin, *J. Power Sources*, 71,268 (1998).
6. A. Smirnova, G. Crosbie, K.Ellwood, *J. Electrochem. Soc.*, **148**, 610 (2001).
7. Y. Funahashi, T. Shimamori, T. Suzuki,Y. Fujishiro, M. Awano, *J. Power Sources*, **163** (2), 731 (2006).
8. X. Zhou, J. Ma, F. Deng, G. Meng, X. Liu, *J. Power Sources*, **279**, 162 (2006).
9. S. Livermore, J. Cotton, R. Ormerod, *J. Power Sources*, **86**, 411 (2000).
10. U. B. Pal, S. Gopalan, W. Gong, FY 2004 Annual Report, Office of Fossil Energy Fuel Cell Program, 262 (2004).
11. N.M. Sammes, Y. Du, R. Bove, *J. Power Sources*, **145**, 428 (2005).
12. J. W. Fergus, *J. Power Sources*, **162**, 30 (2006).
13. M. Mogensen, S. Primdahl, M.J. Jorgensen, C. Bagger, *J. Electroceram.*, **5**, 141 (2000).
14. M. Mogensen, S. Skaarup, *Solid State Ionics*, **86**, 1151 (1996).
15. V.V. Kharton, E.V. Tsipis, I.P. Marozau, A.P. Viskup, J.R. Frade and J.T.S. Irvine, *Solid State Ionics*, **178(1-2)**, 101 (2007).
16. J.Y. Yi, G. M. Choi, *Solid State Ionics*, **175**, 145 (2004).
17. X.J. Chen, Q.L. Liu, K.A. Khor and S.H. Chan, *J. Power Sources*, **165(1)**, 34 (2007).
18. T. Suzuki, T. Yamaguchi, Y. Fujishiro and M. Awano, *J. Power Sources*, **160**, 73 (2006).
19. COMSOL 3.2 User Guide, COMSOL Inc., Burlington, MA, (2006).
20. C. Xia and M. Liu, *Solid State Ionics*, **423**, 152 (2002).
21. X. Xue, J. Tang, N. M. Sammes and Y. Du, *J. Power Sources*, **142**, 211 (2005)
22. Bance, P., Brandon, N.P. Girvan, P. B., Holbeche, P., O'Dea, S. and Steele, B. C. H., *J. Power Sources* **131**, 86 (2004).

PERFORMANCE OF THE GEN 3.1 LIQUID TIN ANODE SOFC ON DIRECT JP-8 FUEL

M. T. Koslowske, W. A. McPhee, L. S. Bateman, M. J. Slaney, J. Bentley and T. T. Tao,
CellTech Power, LLC; Westborough, Massachusetts 01581, USA

ABSTRACT

CellTech Power's Liquid Tin Anode SOFC (LTA-SOFC) has been demonstrated to run directly on gaseous, liquid and solid carbonaceous fuels without fuel processing, reforming or sulfur removal. Examples of complex fuels previously run include natural gas, carbon, biomass, recycled plastic and JP-8 logistic fuel. For these complex fuels, the LTA-SOFC system is projected to be more efficient (30% to 50%) and dramatically simpler than competing systems. A limiting factor in previous cell designs has been low power density for direct fuel conversion, 40 mW.cm^{-2} operating on JP-8 compared to conventional nickel anode based SOFC's operating on hydrogen. Under a DARPA/Army Direct JP-8 conversion program, a high power density Gen 3.1 LTA-SOFC cell has been developed incorporating several new features in design, components and materials. Experimental test results with JP-8 fuel have exceeded key performance goals, including specific power greater than 100mW.cm^{-2}. Coupled with a 4X reduction in weight and volume over previous cells developed in 2006, the Gen 3.1 LTA-SOFC is within the design requirements for military and commercial portable power applications. This paper will discuss some of the design challenges and present single cell and multiple cell performance results.

INTRODUCTION

The Liquid Tin Anode - Solid Oxide Fuel Cell (LTA-SOFC) is based on the unique physical and chemical properties of tin. The use of a liquid tin anode allows CellTech Power SOFC systems to be fuel flexible without auxiliary fuel processing to remove sulfur and reform fuels. It is this aspect which makes LTA-SOFC attractive for generating power from common fuels such as natural gas, propane, diesel, kerosene, gasoline, biodiesel, ethanol, Fischer-Tropes fuel, biomass and coal [1]. The practical application of the LTA-SOFC has been demonstrated on a multitude of these carbonaceous fuels at a single cell, stack and system level over the history of the company [2, 3]. Two prototype standalone systems rated at 1 kW were demonstrated for two thousand plus hours each on natural gas and hydrogen. It is important to acknowledge that the LTA-SOFC can operate with high efficiency on hydrogen and is well suited as a flexible power generation device when the hydrogen economy arrives.

Currently, LTA-SOFC development is targeted for portable power applications using logistic fuels. Liquid petroleum based logistic fuels have a significant volumetric and gravimetric energy density advantage compared to other fuels which makes them highly desirable for mobile/portable applications. Increasing deployment of electronic devices on the battlefield has pushed the US military portable power demands to an unprecedented level. The US Department of Defense is the largest single consumer of oil in the world, with the majority of this oil being consumed as jet fuel, JP-8 [4]. The DOD operates under Directive 4140.25 which gives precedence to the use of JP-8 fuel for all military forward deployed equipment [5]. A renewable or rechargeable power supply compatible with the current logistic fuel is required to simplify supply chain management. Fuel cells in general offer significant technological advantages for portable power application, with efficiency and reliability quoted as the most advantageous qualities.

The use of the LTA-SOFC allows direct utilization of JP-8 fuel at the anode, eliminating the need for reforming and desulfurization. JP-8 has a typical sulfur content of up to 3,000 ppm which is severely detrimental to the catalytic activity of conventional fuel cell anodes. The tin-sulfur phase diagram at 1000°C shows reasonable sulfur solubility in tin, up to 2.5 wt%, with a mixture of liquid tin and tin-sulfide existing up to 20 wt% sulfur [6]. The solubility of sulfur in tin allows use of logistic fuels with high sulfur contents which are detrimental to conventional SOFC anodes.

In the past, the technological barrier for LTA-SOFC running on JP-8 has been low power density, 40 mW.cm^{-2} for the Gen 3.0 cell design. A major component of the total loss was identified as mass diffusion limitations supplying fuel to the liquid tin anode. This key area was addressed through the development of a highly porous material for the anode to increase mass transport to the liquid tin. This design change was incorporated in the Gen3.1 design that resulted in a power density increase to 100 mW.cm^{-2}.

Tin Anode Chemistry and Electrochemistry

Tin is a metallic element with unsaturated p-electrons. It has a very low melting point (mp 232°C) but an exceptionally high boiling point (bp 2,602°C). The chemical reaction of tin with oxygen is exothermic and spontaneous as shown in Equation 1. The electrochemical reaction between tin and air using oxygen transporting membranes, such as the SOFC ceramic YSZ electrolyte, at 1,000°C is given in Equation 2. At the anode, tin combines with oxygen ions to form tin oxide and produces an Open Circuit Voltage (OCV) of 0.78 volt with ambient air at the cathode. In this instance, tin is a reactive and consumable anode similar to a battery anode.

$$Sn\ (l) + O_2\ (g) = SnO_2 \qquad \Delta G = -311\ kJ \qquad T = 1000°C \quad (1)$$

$$Sn\ (l) + 2O(2e\text{-}) = SnO_2 + 4e\text{-} \qquad OCV = 0.78V \qquad\qquad (2)$$

Representative I-V and I-W performance curves of the LTA-SOFC are illustrated in Figure 1. These data are based on experimental results of a single tubular Gen 2.0 cell with electrolyte area of 14 cm^2 on hydrogen and ambient air at 1,000° C. The LTA-SOFC behaves similarly as a traditional SOFC related to its electrode polarizations and internal electric resistances. A fully charged LTA-SOFC produces a cell voltage of 1.06-1.08V, identical to Ni-YSZ SOFCs. However, unlike other fuel cells, LTA-SOFC has a threshold potential of 0.78 V (1,000°C, air cathode). Continuous operation below this threshold leads to a buildup of solid tin oxide. By manipulating operational voltage, LTA-SOFC can be operated either in fuel cell mode for long continuous operation or in battery mode, which can be used for emergency power or during short power bursts.

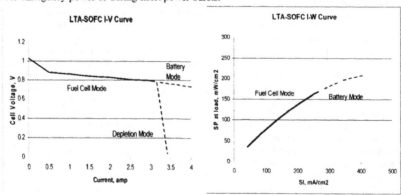

Figure 1. LTA-SOFC I-V-W curves at 1,000°C for a Gen 2.0 button cell showing continued operation is possible in the tin oxide formation regime.

To operate the LTA-SOFC in the fuel cell mode and provide stable power, the tin oxide formation must be controlled. The fuel introduced to the anode is used to reduce the tin oxide back to tin. Equations 3 to 6 show reduction pathways for SnO_2 back to Sn with the base constituents of hydrocarbon fuels such as JP-8, namely hydrogen and carbon. Carbon monoxide (CO) is also present as an intermediate product at standard pressure and 1,000°C. All these reduction reactions are spontaneous as shown by their negative free energies.

$$SnO_2 + 2H_2(g) = Sn(l) + 2H_2O(g) \qquad \Delta G = -44 \text{ kJ} \qquad T = 1000°C \quad (3)$$

$$SnO_2 + 2C = Sn(l) + 2CO(g) \qquad \Delta G = -137 \text{ kJ} \qquad T = 1000°C \quad (4)$$

$$SnO_2 + C = Sn(l) + CO_2(g) \qquad \Delta G = -85 \text{ kJ} \qquad T = 1000°C \quad (5)$$

$$SnO_2 + 2CO(g) = Sn(l) + 2CO_2(g) \qquad \Delta G = -33 \text{ kJ} \qquad T = 1000°C \quad (6)$$

The equations above indicate that in the presence of fuel molecules, such as hydrogen, carbon, CO or any intermediate hydrocarbon species that is readily thermally decomposed in situ, formation of tin oxide can be reversed. With this understanding and the addition of Equation 2, the net result is that electricity is being produced in a way as if a direct oxidation of the fuel molecules occurs. In other word, LTA-SOFC acts like a "direct fuel" or "direct carbon" conversion fuel cell. During operation, the tin anode acts as a shuttle mechanism for transport of oxygen ions from the electrolyte to the fuel. The tin has two discreet mechanisms. One is the generation of electrons via the formation of tin oxide and current path. The second is the oxidation of fuel and the subsequent reduction of the tin oxide. In effect, these functions of the tin may be decoupled in the design of larger stationary power units. Thermodynamic calculations predict spontaneous reduction of tin oxide by hydrogen and carbon above 500°C which provides a large window of protection for the liquid tin during thermal cycling.

Experimental validation of the ability of LTA-SOFC operation on solid fuels was established in 1999 when the first generation design was operated on coal and carbon. Efficient operation on carbon and coal has been validated through single cell tests in subsequent cell designs such as that shown in Figure 2. In previous tests, 60% carbon fuel efficiency has been measured for a single cell compared to a projected fuel efficiency of 70% for LTA-SOFC.

Cell Design

Figure 2 shows the progression of cell design from Gen 3.0 to Gen 3.1 for portable power applications. Gen 3.0 was developed under the DARPA MISER program to utilize battlefield plastic waste to produce usable power. The requirement for smaller and lighter systems under a DARPA/ARMY direct JP-8 program resulted in the Gen 3.1 cell design. Improvements in material tolerances and the understanding of the liquid tin requirements lead to a reduction of weight and volume of four times. Along with this physical scaling, the specific power was increased by a factor of three from 40 mW.cm^{-2} to 120 mW.cm^{-2} on JP-8. The primary cause for the power increase was the design of an open porous membrane to both contain the liquid tin and allow transport of the fuel molecules. The wetting behavior of the liquid tin allows this. Although published data varies, the wetting behavior of tin has been experimentally determined by the authors to be poor in a reducing atmosphere. The poor wetting allows liquid tin at 1000°C to be contained in a highly porous structure, thus allowing favorable mass diffusion of the fuel to the liquid tin anode.

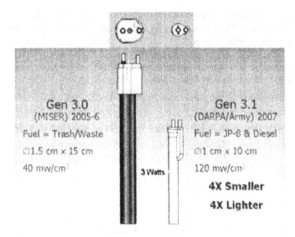

Figure 2. Comparison of 3 Watt Gen 3.0 and Gen 3.1 LTA-SOFC single cells

EXPERIMENTAL

Single Cell Construction

The Gen 3.1 cell is constructed around a closed end 8mol% yttria stabilized zirconia (8YSZ) electrolyte tube with nominal thickness between 160 and 200 microns. The 8YSZ powder was sourced from Tosoh Corporation as TZ-8Y. The electrolyte tubes were sintered at 1600°C and then helium leak checked. Cathodes and cathode current collection components are produced in-house. The single cell assembly starts with a half cell consisting of electrolyte, cathode and cathode current collection. This assembly is then bonded with ceramic glue to an alumina cap which seals and separates the cathode from the anode. An alumina porous separator that encapsulates the electrolyte is then attached to the alumina cap. The porous separator has 65% porosity with pore diameters <200 μm. The porous separator is produced in house by a propriety manufacturing process. Tin from Surepure Chemetals Inc. with a purity of 99.5% was used for the anode.

The single cell is then sealed within an alumina closed end tube for testing. Delivery and exhaust of the fuel and oxidant are provided by alumina tubes that go through the sealed top of the process tube.

Single Cell Test Procedure

The complete single cell test assembly is placed in a Carbolite vertical tube furnace. The cells were typically heated at 120°C/hr, with a dwell at 400°C to load tin into the anode. Dry air was supplied to the cathode from an Atlas Copco GX7FF compressor. The air flow rate was kept constant (300 cc.min^{-1}) using a 0-500 ml rotameter with a manually adjustable needle valve. Hydrogen, Spec Air – Specialty Gases industrial grade hydrogen (99.99% purity), was delivered to the cell during heating and benchmark testing at a flow rate of 300 cc.min^{-1} through a MKS mass flow controller. Current collection was protected using a cover gas mixture of 95% Argon – 5% Hydrogen, Spec Air – Specialty Gases. A Chroma load box and MKS mass flow controllers were controlled using a Labview based test program through a National Instruments Field Point DAQ.

The JP-8 fuel was supplied by Natick Soldier Center, Natick, MA and contained a sulfur concentration of 1400 ppm. The fuel was delivered to the fuel chamber at flow rates from 5 to100 μl.min^{-1} by an IVEK piston pump (model 102132). The JP-8 flow rate was calibrated using a stopwatch and balance. This calibration was checked throughout experimentation by recording the weight of JP-8 consumed as a function of time. JP-8, without any form of fuel processing or sulfur removal, was directly injected into the anode chamber via a needle.

The cells were qualified on hydrogen to confirm performance before switching to JP-8. The cell I-V performance was characterized in a way similar to traditional SOFC. However, because there exists a tin anode Nernst potential threshold (0.78V at 1,000C in air), cells operating at or below this threshold will form solid tin oxide, resulting in tin anode consumption. For easy comparison between tests, a cell maximum current was defined as the current that the cell was able to provide continuously at a constant cell voltage (less than 0.002 volt decline) over 5 minutes.

Fuel Utilization

Fuel utilization calculation for a simplistic hydrogen fuel is not easily applied to a complex hydrocarbon. For a hydrocarbon the electrochemical reaction will utilize both the carbon and hydrogen via the following half reactions, equations 7 to 9, etc, at the anode where each carbon either release 2 electrons (forming CO) or 4 electrons (forming CO2).

$$H_2 + O^= \rightarrow H_2O + 2e^- \tag{7}$$

$$C + O^= \rightarrow CO + 2e^- \tag{8}$$

$$C + 2O^= \rightarrow CO_2 + 4e^- \tag{9}$$

For a hydrocarbon of composition C_nH_m , the maximum number of electrons released during the electrochemical reaction at the anode can be expressed as

$$\text{Max Number of } e^- = 4n + 1m \tag{10}$$

Using a simplistic representation of JP-8 as $C_{16}H_{34}$, the maximum amount of charge for 1 mole of JP-8 can be determined in equations 11 and 12 and used to predict the available electrical potential as a function of fuel flow, equation 13.

$$\text{Max Number of } e^- = 4 \times 16 + 1 \times 34 = 98 \tag{11}$$

$$C_{16}H_{34(l)}\left(mol \cdot min^{-1} \cdot A^{-1}\right) = \frac{1}{96,487}\left(\frac{C/s}{C}\right) \times \frac{1}{98}(g\ mol) \times 60\left(\frac{s}{min}\right) = 6.34 \times 10^{-6} \tag{12}$$

$$C_{16}H_{34(l)}\left(\mu l \cdot min^{-1} \cdot A^{-1}\right) = 6.34 \times 10^{-6}\left(mol \cdot min^{-1} \cdot A^{-1}\right) \times 226.5\left(g \cdot mol^{-1}\right) \times \frac{1000}{0.8}\left(\mu l \cdot g^{-1}\right) = 1.796 \tag{13}$$

In this prediction, the fuel flow rate used for experimentation divided by 1.796 $\mu l.min^{-1}A^{-1}$ gives the maximum current that can be drawn from the cell. At a fuel flow rate of 10 $\mu l.min^{-1}$ this calculation suggests the maximum current would be 5.57 A at 100% fuel utilization. This calculation however, is based on an assumption that the chemical formula and therefore the molecular weight of JP-8 are known. In reality the molecular weight is not known and there are hundreds of hydrocarbons present in a typical sample of native JP-8 fuel and as intermediate pyrolyzation species.

Fuel Efficiency

Since the fuel utilization of complex hydrocarbons such as JP-8 is ill defined, the fuel efficiency was used as a performance metric. The LHV for JP-8 is typically reported as 43,190 kJ.kg^{-1}. Therefore the thermal power input from JP-8 at a known flow rate of 0.133 mg/s can be calculated as shown in Equation 14.

$$\text{Thermal Input Power}(W) = X \times LHV\left(kg \cdot s^{-1} \cdot J \cdot kg^{-1}\right) = 0.000000133 \times 43,190,000 = 5.744(W) \quad (14)$$
$$\text{where } X = \text{JP-8 flow rate - 133.33 } \mu g.s^{-1}$$

The fuel efficiency is then calculated in Equation 15 from the thermal input power and the electric power output at load. Simple stated, the closer the power output of a cell or system to the thermal input power of the fuel the higher the efficiency.

$$\text{Fuel Efficiency (\%)} = \frac{\text{Power Output at load (W)}}{\text{Thermal Input Power (W)}} \quad (15)$$

RESULTS

Single Cell Performance

At the time of publication over 50 single Gen3.1 cells have been tested. Performance of a Gen 3.1 single cell on JP-8 is shown in Figure 3 as a typical I-V curve with characteristic polarization at low current densities and a linear decrease in voltage at higher current densities. Its OCV was about 1.03-1.05 V. The maximum power density measured was 120 mW.cm^{-2} at 50 μl.min^{-1} flow rate, with a current density of 220 mA.cm^{-2}. This resulted in a stable 3.3 watt cell. In comparison, the maximum performance on hydrogen (300 cc.min^{-1}) was 153 mW.cm^{-2}, power density, and 316 mA.cm^{-2}, current density, yielding 4.8 watts.

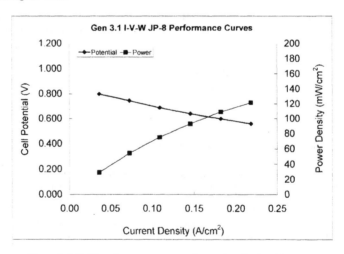

Figure 3. I-V-W performance curves of a Gen3.1 single cell on JP-8.

The I-W power density comparison between hydrogen and JP-8 for a single cell is plotted in Figure 4 for flow rates of 300cc.min^{-1} and 50 µl.min^{-1} respectively. Typically the cell performance achieved on JP-8 was 70-80% of the hydrogen performance.

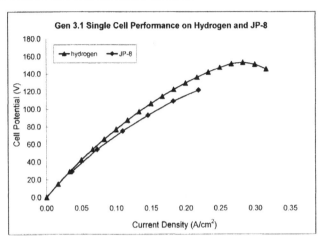

Figure 4. Gen 3.1 single cell performance comparisons between hydrogen and JP-8.

Long term stability test data for direct JP-8 conversion in a Gen 3.1 cell is presented in Fig 5. In the first 100 hrs of the test, there was a slight increase in the performance in the power and potential over this time. JP-8 flow rate was fixed at 10 µl.min^{-1}. The efficiency during the first 100 hours was calculated to be 17.4%. Overall cell performance was stable and no degradation was observed, despite occasional smoke exited the exhaust.

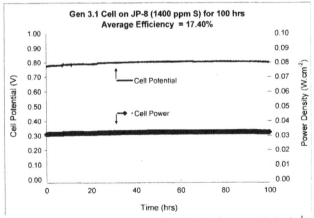

Figure 5. Gen 3.1 single cell longevity test. JP-8 flow rate 10 µl.min^{-1}

After the 100 hour performance goal was achieved, the current load of cell was increased incrementally to give a maximum fuel efficiency up to 41% at 76 mW.cm^{-2} for more than one hour as shown in Fig 6. There was a slight degradation in cell potential which suggested this was near the threshold for stable operation. Then, the cell was operated for 10 hours continuously at above 30% efficiency. There was no significant degradation in the cell performance during this period. The fuel efficiency was calculated in the way of traditional electric generators using Equation 15.

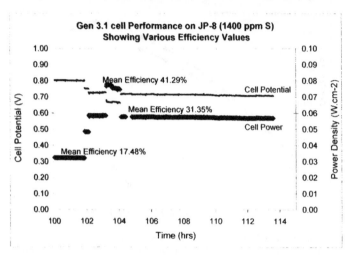

Figure 6. Gen 3.1 cell fuel performance and fuel efficiency values for a direct JP-8 conversion. More than 40% fuel efficiency was demonstrated.

Gen 3.1 Bundle / Stack Results

The performance, power and potential, of a four cell bundle on JP-8 is shown in Figure 7. A bundle consists of four single cells in the same anode chamber sharing a common fuel source, connected electrically, but having individual air delivery. In this instance, the cells were connected in parallel. The bundle yielded a maximum power of 9.2 watts at fifteen amps of current. JP-8 was directly injected into the anode chamber at a flow rate of 100 μl.min^{-1} without any fuel process or reforming.

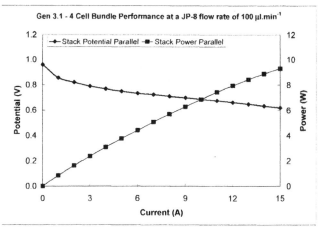

Figure 7. Gen 3.1 four cell bundle I-V-W performance on JP-8 @ 100 ml.min⁻¹.

DISCUSSION

Performance of the Gen 3.1 LTA-SOFC on complex fuels, namely the US military logistic fuel JP-8, shows encouraging results for development in portable power applications. One material development of interest that has directly contributed to the increased performance is that of the porous separator. The porous separator has two main functions. The first function as illustrated in figure 8 is concerned with the containment of the liquid tin anode in contact with the electrolyte at a uniform tin thickness and without separation in the tin column or leakage of the tin. The second function is to provide an open diffusion path for fuel molecules to reach the liquid tin surface. The minimum dimensions for the tin thickness and pore structure can be calculated using a capillary model and head pressure equations. The result of a thin tin thickness and low head pressure is a break or gap in the continuity of the tin column caused by surface tension effects. Since the liquid tin also provides the current path, a break or separation of tin should be avoided and this would severely reduce the active area of the cell. A nominal tin thickness of 500 microns is sufficient to insure a uniform tin column. The pressure head of the tin at the base of the cell can also be calculated to predict the maximum connected pore channel diameter that will safely contain the liquid tin. It has been determined and verified that a connected pore channel less than 150 microns is sufficient to contain the tin anode in Gen 3.1 cells.

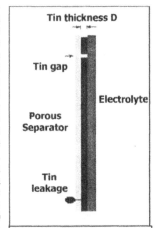

Figure 8. Schematic of porous separator function in LTA-SOFC.

The optimization of the porous separator has resulted in higher sustainable power at all flow rates, which corresponds to higher efficiencies. Table I summarizes the JP-8 efficiency data for Gen 3.1 cells. The 100 hr durability test had an average efficiency of 17.4%. An average efficiency for the 10 hr period at the end of cell life was 31.66%. A maximum efficiency of 41.29% was measured. The data suggests that >30% efficiency for 100 hr is achievable using the current cell geometry. This compares favorably against the requirements for military portable power generators. The efficiency is a more useful number since it takes into account the resistive losses during operation.

Table 1. Experimentation showed that efficiencies up to 41% were sustainable for a minimum of an hour with a JP-8 flow rate of 10 μl.min^{-1}.

Cell Current (A)	Fuel Utilization (%)	Power Density (mW.cm^{-2})	Fuel Efficiency (%)
1.247	22.38	32	17.48
1.995	35.82	48	26.18
2.490	44.70	58	31.66
3.536	63.48	76	41.29

The difference between the porous separators from Gen 2.0-3.0 to Gen 3.1 is shown in Figure 9. The layer structure of the LTA-SOFC with a finer and less porous mixed ionic-electronic conductor material. The mean pore diameter is on the order of microns. In contrast the porous separator developed for JP-8 testing using a structural ceramic shows large interconnected porosity with a mean diameter > 100 microns.

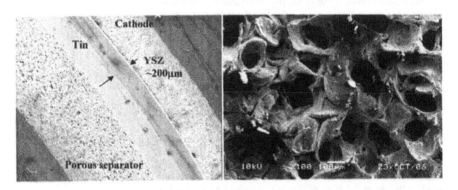

Figure 9. Porous separator microstructure for Gen 2.0-3.0 cell (left) and Gen 3.1 (right) showing evolution of material porosity to large 100 micron channels.

Confirmation of the porous separator's contribution is shown in Figure 10, comparing single cell performance data on JP-8 between the Gen 3.0 and Gen 3.1 cells. The increase in cell performance is mainly attributed to reduced anode polarization.

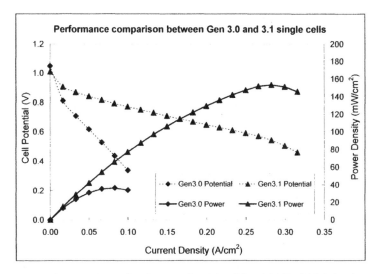

Figure 10. Comparison between Gen 3.0 and Gen 3.1 LTA-SOFC showing
marked improvement in cell performance.

ONGOING WORK

Development of a LTA-SOFC stack is continuing for 10 W, 60 W and up to 250 W systems based on logistic fuels. In other areas, interest in the direct fuel conversion has lead to testing of the Gen 3.1 cell on Bio-derived charcoal and coal. The unique properties of the liquid tin are advantageous for the direct carbon conversion made possible by the decoupling of the electrical generation and tin reduction. Power plant scale systems could efficiently convert coal for base load power.

CONCLUSIONS

The Liquid Tin Anode SOFC has a unique anode structure which results in the ability to directly convert any carbon containing fuel to usable power. Currently cell designs are being optimized for portable power applications. Power plant electricity production using coal and biomass is interesting due to the high conversion efficiency of the LTA-SOFC. The following conclusions were drawn from this work:

1. The single cell maximum power density running directly on JP-8 was 120 mW.cm^{-2}
2. The fuel efficiency over 40% was measured on unreformed JP-8.
3. A four cell bundle was able to produce 9 W on JP-8.
4. An improved porous separator with larger pore size was the key design change that allowed improvement in gravimetric and volumetric densities compared to Gen3.0 cells.
5. The Gen 3.1 porous separator resulted in a 3x performance improvement.
6. No reforming or desulfurization of the JP-8 (sulfur 1,400ppm) was required. Gen 3.1 would be suitable for direct fuel conversion of common fuels, including gaseous, liquid and solid fuels such as NG, propane, gasoline, alcohols, biomass and potentially coal. One particular example was a sub-kilowatt, mobile/portable, JP-8 operated, electricity generation device for military field battery chargers.

ACKNOWLEDGEMENTS
Development of Gen 3.0 was partially supported under DARPA Contract W911QY-04-2-003;
Development of Gen 3.1 was partially supported under DARPA/ARMY Contract W911NF-07-C-0032

REFERENCES
1. J. Bentley, T, Tao, *Liquid Tin Direct Fuel Cell for JP-8, Coal and Biomass Applications*, DOD 6th Annual Logistic Fuel Processing Conference, May 2006.
2. T. Tao, L. Bateman, J. Bentley and M. Slaney, *Liquid Tin Anode Solid Oxide Fuel Cell for Direct Carbonaceous Fuel Conversion*, 2006 Fuel Cell Seminar. p. 198
3. T. Tao, *Introduction of Liquid Anode/Solid Oxide Electrolyte Fuel Cell and its Direct Energy Conversion using Waste Plastics*, in SOFC-IX, S.C. Singhal and J. Mizusaki, editor, March 2005, Vol 1, p 353
4. S. Karbuz, Energy Bulletin, Feb (2007)
5. Wolfowitz, P., 2004, *Department of Defense Directive 4140.25 – DoD Management Policy for Energy Commodities and Related Services*, Department of Defense, Washington, pp.1-10.
6. R. S. Roth, T. Negas, and L. P. Cook, Editors, Phase Diagrams for Ceramists, Vol V, The American Ceramic Society, (USA).

EFFECT OF INTERCONNECT CREEP ON LONG-TERM PERFORMANCE OF SOFC OF ONE
CELL STACKS

W.N. Liu, X. Sun, and M.A. Khaleel
Pacific Northwest National Laboratory
Richland, WA 99354

ABSTRACT
        High temperature ferritic alloys are potential candidates as interconnect (IC) materials and
spacers due to their low cost and CTE compatibility with other SOFC components for most of the solid
oxide fuel cells (SOFC) under development in the SECA program. Possible creep deformation of IC
under the typical cell operating temperature should not be neglected. In this paper, the effects of
interconnect creep behavior on stack geometry change and stress redistribution of different cell
components are predicted and summarized. The goal of the study is to investigate the performance of
the fuel cell stack by obtaining the fuel and air channel geometry changes due to creep of the ferritic
stainless steel interconnect, therefore indicating possible SOFC performance change under long term
operations. IC creep models were incorporated into SOFC-MP and Mentat FC, and finite element
analyses were performed to quantify the deformed configuration of the SOFC stack under the long
term steady state operating temperature. It is found that creep behavior of the ferritic stainless steel IC
contributes to narrowing of both the fuel and the air flow channels. In addition, stress re-distribution
of the cell components suggests the need for a compliant sealing material that also relaxes at operating
temperature.

INTRODUCTION
        Interconnects in solid oxide fuel cells (SOFCs) provide cell to cell electrical connection, and
also serve as gas separator for the separation of the fuel (anode) from the oxidant (cathode). In order to
perform their intended functionalities, interconnects must demonstrate or possess the following
materials characteristics: (1) Excellent electrical conductivity; (2) Corrosion resistance in oxidizing
and reducing environment; (3) Corrosion resistance in bi-polar exposure condition; (4) Matched
thermal expansion coefficient (CTE) with other SOFC components; (5) Good thermal conductivity for
thermal management; (6) Bulk and interface stability with electrodes; (7) Low/negligible solubility for
hydrogen, carbon and oxygen; and (8) Low cost and easy to fabricate .
        Recently, the interconnect materials development has been mostly focusing on ferritic stainless
steels. Compared to chromium-based alloys, iron-based alloys have advantages in terms of high
ductility, good workability and low cost [1]. By far, iron-based alloys, especially Cr-Fe based alloy,
e.g. Crofer 22 APU, are the most attractive metallic interconnect material for SOFCs [2, 3].
        Creep deformation becomes relevant for a material when the operating temperature is near or
exceeding half of its melting temperature (in degrees of Kelvin) [4, 5, 6]. The operating temperatures
for most of the solid oxide fuel cells (SOFC) under development are around 1073°K. Since the melting
temperature of most stainless steel is around 1800°K, creep deformation of IC under the typical cell
operating temperature should not be neglected.
        The temperature differential between the initial, stress-free fabrication temperature and SOFC
operating temperature will cause stresses and deformation of various cell components. Under long term
operation, the stresses and deformation are expected to relax and re-distribute due to the anticipated
creep behavior of the interconnect material.
        In this paper, the effects of interconnect creep behavior on one/multi-cells stack geometry
changes and stress redistributions in various cell components are predicted and summarized. The goal
of the study is to investigate the long term performance of the fuel cell stack by obtaining the fuel and

air channel geometry changes due to the creep behavior of the ferritic stainless steel interconnect. Finite element electrochemical analyses were first performed to determine the steady-state operating temperature profile using SOFC-MP. The temperature differential between the initial, stress-free fabrication temperature and SOFC operating temperature will cause stresses and deformation of various cell components. IC creep laws have been incorporated into SOFC-MP and Mentat FC. and finite element analyses were performed to quantify the deformed configuration of the SOFC under the long term steady state operating temperature. It is found that creep behavior of the ferritic stainless steel IC contributes to narrowing of both the fuel and the air flow channels. Stresses relaxed in IC due to IC creep will be transferred to other parts and cause the re-distribution of stresses in SOFC. Stress re-distribution of the cell components suggests the need for a compliant sealing material that relaxes at operating temperature.

TEMPERATURE/TIME DEPENDENT PROPERTIES OF MATERIAL IN SOFC

The mechanical properties of a commonly used high temperature ferritic alloy SS430 is used in the current numerical analyses. The temperature dependent Young's modulus and CTE for SS430 are shown in Figures 1 and 2 [7]. Since the creep behavior of SS430 is not available in the open literature, the complete creep behaviors of Fecralloy stainless steel under various initial stresses are used, see Figure 3. It should be mentioned that even though the exact creep rate vs. temperature curves could be different from SS430 to Fecralloy, the trend of the creep curves should be similar between the two materials with increasing temperature.

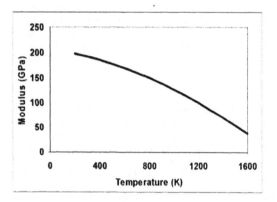

Figure 1 Temperature dependent modulus of SS 430

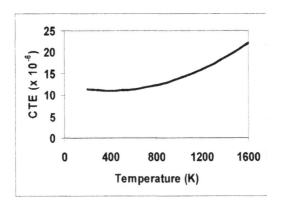

Figure 2 Temperature dependent CTE of SS430

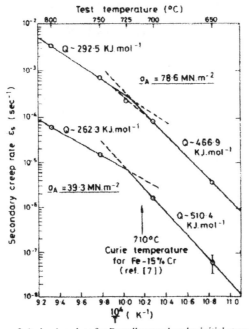

Figure 3 Arrhenius plots for Fecralloy steel under initial stress [4]

The creep rate shown in Figure 3 may be expressed as [4]
For $\sigma < 100 MPa$

$$\dot{\varepsilon}_c = \begin{cases} 1.72\sigma^{5.5}e^{-277400/RT} & \text{for} \quad T \geq 725^\circ C \\ 1.65\times10^{11}\sigma^{5.8}e^{-488700/RT} & \text{for} \quad T \leq 710^\circ C \end{cases} \quad (1)$$

and for $\sigma > 100 MPa$

$$\dot{\varepsilon}_c = \begin{cases} >10^{-3} & \text{for} \quad T \geq 725^\circ C \\ 28.2\times\sigma^{10.5}e^{-488700/RT} & \text{for} \quad T \leq 710^\circ C \end{cases} \quad (2)$$

This creep law has been implemented in Mentat-FC using a user-defined material subroutine.

GEOMETRY OF ONE/THREE CELLS STACK AND FINITE ELEMENT MODEL

The initial geometry of a one-cell stack is illustrated in Figures 4. Its in-plane dimensions are 157.9mm by 149.5mm, and the total thickness is 5mm. This is the default geometry in SOFC-MP with cross flow.

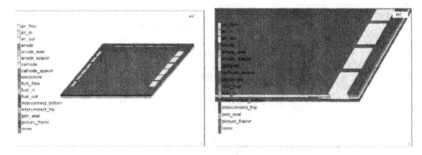

Figure 4 Illustration of one-cell stack

Since the SOFC components are considered to be stress free at its assembly temperature of 800°C, different degrees of thermal stresses will be generated at the steady state operating temperature due to the CTE mismatch of various cell components. Figure 5 shows the finite element model used in the thermal-mechanical analysis.

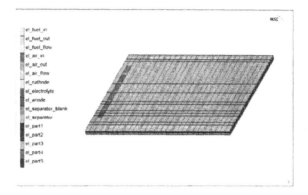

Figure 5 Finite element model used in the thermal-mechanical analysis

To obtain the steady-state operating temperature profile, SOFC-MP was used to perform the electro-chemical-thermal analyses. Figure 6 depicts the steady state IC temperature distributions on the surfaces.

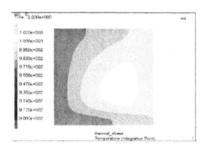

Figure 6 Steady state temperature distributions

NUMERICAL RESULTS AND DISCUSSIONS

Figures 7 and 8 show the deformed configurations of the fuel and air flow channels of one cell stack at $t=0.01h$, respectively, with deformation magnification factor of 100. From Figure 7(b), it can be seen that the height of the flow channel narrows down with the temperature differential from the stress-free temperature to the steady-state temperature due to the mismatch and creep behavior of IC materials. This channel geometry change after reaching operating condition is caused by the CTE mismatches of various cell components such as anode, seal and interconnects.

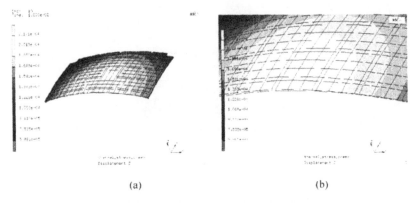

(a)   (b)

Figure 7 Deformed configuration of fuel flow channel

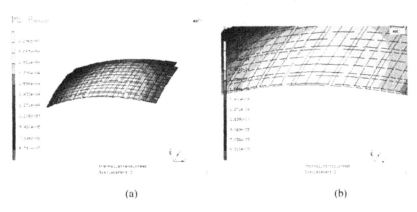

(a)   (b)

Figure 8 Deformed configuration of air flow channel

Figures 9 and 10 illustrate the configurations of deformed IC and PEN assembly at $0.01h$ and $1000h$, respectively. At the end of 0.01hour, the z-displacement of the bottom IC is the highest at the center of the cell, while the z-displacement of the PEN is the least at the same location. Therefore, the air flow channel is reduced and fuel flow channel is expanded due to the initial CTE mismatch. At the end of $1000h$ (Figure 10), however, more IC creep behavior is observed: at the center of the cell, the z-displacement of the bottom IC is the largest, but the top IC is the least. This means that both the air flow channel and fuel flow channel are reduced due to the creep behavior of the IC under the operating temperature.

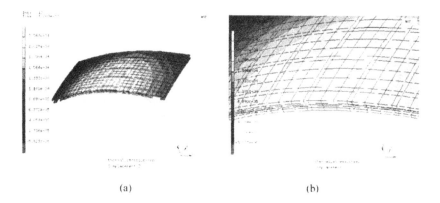

(a)   (b)

Figure 9 Deformed IC and PEN at 0.01hr

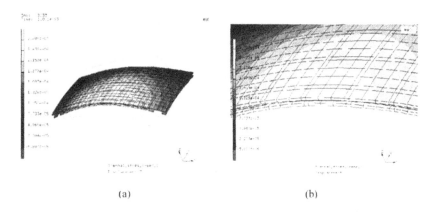

(a)   (b)

Figure 10 Deformed IC and PEN at 1000hr

Figure 11 shows the changes of the air and fuel flow channel widths at the center of the cell within the first hour of SOFC operation after cooling down from initial stress-free condition. Initially, CTE mismatch reduces the height of air flow channel but expands the height of fuel flow channel slightly. Subsequently, the creep behavior of IC begins to contribute to the height reductions of both air and fuel flow channels. The effect of IC creep on the changes of flow channels is shown in Figure 12 for the first one-thousand hour operating period.

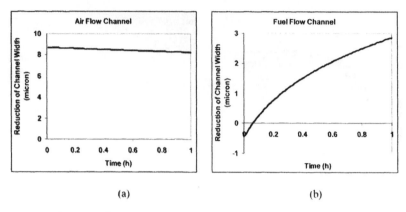

(a)   (b)

Figure 11 Effect of creep of IC on flow channels within initial one hour
(a) air flow channel, (b) fuel flow channel

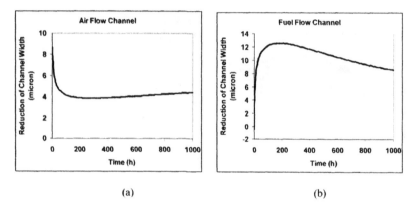

(a)   (b)

Figure 12 Effect of creep of IC on flow channels within 1000 hour
(a) air flow channel, (b) fuel flow channel

Figures 13 and 14 illustrate the effects of IC creep on stress evolutions in various cell components in the first 68 hours of operation. It may be observed that the stresses in both the bottom and the top ICs are relaxed due to creep, the stresses in the PEN structure are also reduced due to the IC stress relaxation. However, PEN seal stresses remain at a relatively high level during operation without considering its creep behaviors. The maximum Von Mises stresses in various components are listed in Table at three time points.

Table 1 Maximum Von Mises stresses in various components (MPa)

Time (h)	Top IC	Bottom IC	Anode	Cathode	Electrolyte
0.01	31.2	47.2	61.6	52.6	61.6
68	7.2	7.8	55.3	49.5	55.3
1000	4.8	5.0	54.3	49.1	54.3

Time (h)	PEN seal	Picture frame	Anode spacer	Anode seal	Cathode spacer
0.01	52.3	30.1	21.8	18.5	37.5
68	49.5	7.8	7.5	7.3	7.8
1000	49.1	5.8	5.0	5.0	4.9

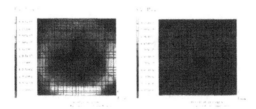

On top Surface

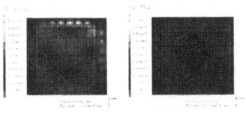

On bottom surface

t = 0.01h   t = 68h

Figure 13 Effect of creep on stress in top IC

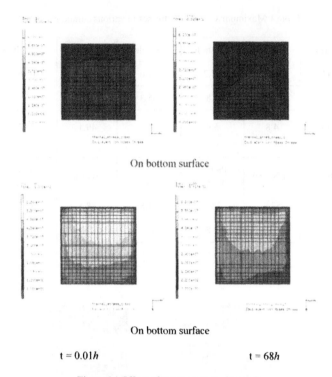

On bottom surface

On bottom surface

t = 0.01*h*   t = 68*h*

Figure 14 Effect of creep on stress in anode

CONCLUSIONS

The effects of interconnect creep on the change of air and fuel flow channels were predicted under the steady state operating condition for a period of 1000 hours. The purpose is to quantify the long term SOFC stack performance by investigating the changes in air and fuel flow paths and the corresponding stress re-distribution among various cell components. The conclusions of the current study are:

1. Mismatch of CTE of the various components in SOFC causes the bending of various components such as interconnects and PEN structure;

2. Non-uniform bending deformations of various cell components cause the change of air and fuel flow path at the initial stage of SOFC operation: the height of air flow channel was reduced, but the height of fuel flow channel was enlarged slightly;

3. Over the course of 1000 hours of operating time, creep behavior of the interconnect contributes to the height reductions of the air fuel flow channels;

4. Effect of the creep of IC on the height of both the air and fuel flow channels is not monotonous due to the different stress levels in the interconnects, which create different stain rates under creep deformation;

5. Creep of the interconnect releases the stress in the interconnect, and reduces the stress level in the PEN structures;
6. Stress in the PEN seal remains at relatively high level without considering seal creep.

REFERENCE

[1] J. W. Fergus, Metallic interconnects for solid oxide fuel cells. *Materials Science and Engineering A (Structural Materials: Properties, Microstructure and Processing)*, 397(1-2), pp. 271-83, 2005.

[2] Z. Yang, G. Xia, and J.W. Stevenson, Mn1.5Co1.5O4 spinel protection Layers on ferritic stainless steels for SOFC interconnect applications. *Electrochemical and Solid-State Letters*, 8(3), pp. 168-70, 2005.

[3] Z. Yang, J.S. Hardy, M.S. Walker, G. Xia, S.P. Simner and J.W. Stevenson, Structure and Conductivity of Thermally Grown Scales on Ferritic Fe-Cr-Mn Steel for SOFC Interconnect Applications, *Journal of The Electrochemical Society*, 151(11), A1825-A1831, 2004.

[4] R.C. Lobb & R.B. Jones, Creep-Rupture Properties of Fecralloy Stainless Steel between 650 and 800°C, *Journal of Nuclear Materials*, 91, pp. 257-264, 1980.

[5] J.E. Benci, D. P. Pope, E. P. George, Creep damage nucleation sites in ferrous alloys, *Materials Science & Engineering A (Structural Materials: Properties, Microstructure and Processing)*, v A103, n 1, pp. 97-102, 1988,

[6] O. R. Arzate and L. Martinez, Creep Cavitation in Type 3 21 Stainless Steel, *Materials Science and Engineering A*, 101, pp.1-6, 1988.

[7] K. I. Johnson, V. N. Korolev, B. J. Koeppel, K. P. Recknagle, M. A. Khaleel, D. Malcolm and Z. Pursell, Finite Element Analysis of Solid Oxide Fuel Cells Using SOFC-MP™ and MSC.Marc/Mentat-FC™, *PNNL report 15154*, Pacific Northwest National Laboratory, June 2005.

EFFECTS OF COMPOSITIONS AND MICROSTRUCTURES OF THIN ANODE LAYER ON THE PERFORMANCE OF HONEYCOMB SOFCs ACCUMULATED WITH MULTI MICRO CHANNEL CELLS

Toshiaki Yamaguchi[1], Sota Shimizu[2], Toshio Suzuki[1], Yoshinobu Fujishiro[1]. Masanobu Awano[1]

[1] National Institute of Advanced Industrial Science and Technology (AIST)
2266-98 Anagahora, Shimoshidami, Moriyama-ku, Nagoya, Aichi, 463-8560 Japan
[2] Fine Ceramics Research Association
Shimoshidami, Moriyama-ku, Nagoya, Aichi, 463-8561 Japan

ABSTRACT
We report herein the effects of the composition and microstructure of thin anode layer on the performance of a cathode-supported honeycomb SOFC accumulated with multi micro channel cells. The volumetric power densities at 0.7 V of the prepared honeycomb SOFC were 0.061 and 0.243 W/cm^3 at 500 and 600 °C, respectively. The power density of our developed cathode-supported SOFC was affected by the anode composition, and the application of the high GDC-content anode greatly improves a rapid I-R drop at low current densities. In addition, we perform electrical simulations of various honeycomb SOFCs. The anode thickness needed to be more than 10 μm in order to avoid the lack of insufficient in-plane conductivity of thin anode layer.

INTRODUCTION
Solid oxide fuel cells (SOFCs) have received a great deal of attention because of the high energy conversion efficiency and the environmental compatibility. Recently, technologies for the miniaturization of the cell size and the integration of the miniaturized cells have been actively investigated in order to reduce operation temperatures, reduce size, and hasten start-up and shut-down operations [1-3]. Among various proposed cell designs, a honeycomb-supported SOFC is considered to be one of the desirable designs for simultaneous achievement of the miniaturized cell size and the integration of the miniaturized cells because of structural advantages, such as the cumulative capacity of multiple cells and the ease with which the cell size and configuration can be controlled [4]. So far, we reported the fabrication technologies for LSM cathode supported honeycomb SOFCs via the extrusion of LSM honeycomb monoliths and the coating of inner walls of multi micro channels using electrolyte and anode slurries [5,6]. The prepared honeycomb SOFC had channel density of about 1000 cpsi (channels per square inch), wall thickness of about 160 μm, the dense electrolyte layer about 10 microns thick, and anode layer about 20 microns thick. The effective electrode area of the honeycomb is estimated to be 40 cm^2 per

1 cm³.

In this study, the effects of thin anode layer on the performance of the cathode-supported honeycomb SOFC were investigated by changing the anode composition and microstructure. Results on the electrical simulation for the honeycomb SOFCs with various anode compositions will also be presented.

EXPERIMENTAL PROCEDURE

In this study, $(La_{0.8}Sr_{0.2})_{0.99}MnO_{3\pm\delta}$ (LSM), $Ce_{0.9}Gd_{0.1}O_{1.95}$ (GDC), Sc_2O_3 doped ZrO_2 (ScSZ) and NiO were used as the raw materials. We prepared two varieties of cathode support for our experiments, a *small tube* of 2 mm in outside diameter and 1 mm in inside diameter and a *honeycomb monolith* with a wall thickness of 0.2 mm and a channel diameter of 2 mm. The mixture of LSM powder and cellulose binder was uniaxially extruded through each die to form the micro cathode tube and honeycomb monolith.

After drying the extruded LSM honeycomb monolith, the channels were coated simultaneously with a ScSZ slurry and co-fired at 1300 °C for 2 h. Subsequently, the honeycomb channels were coated with a 70 wt% NiO- 30 wt% GDC slurry and heated again at 1300 °C for 2 h. The effect of the anode composition on the cell performance was investigated using the tubular cells. The green LSM tube was dip-coated with LSM-GDC slurry to activate the cathode reaction, and then calcined at 1000 °C for 2 h in air. The calcined micro cathode tube was coated with ScSZ electrolyte slurry, and then co-sintered at 1300 °C for 2 h in air to form a dense ScSZ film on the micro support surface. The ScSZ-coated LSM tube was further coated with different NiO-GDC slurries, and then heated at 1300 °C for 2 h in air. The current-voltage measurement was conducted using a Solartron 1260 frequency response analyzer with a 1296 Interface, using wet H_2 (3 % H_2O) and O_2 as fuel and oxidant, respectively.

The current obtained from a cathode-supported honeycomb SOFC model (Fig. 1)

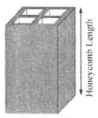

Channel shape : square
Channel number : 2×2
Channel size : 0.64 mm×0.64 mm
Wall thickness : 0.16 mm

Fig. 1. Model structure of a cathode-supported honeycomb SOFC

Table 1. Parameters used in the electrical circuit simulation

cell resistance ($\Omega \cdot cm^2$)	anode layer	cathode honeycomb body
1.0	conductivity: 1500 S/cm thickness: 10~100 μm	conductivity: 100 S/cm thickness: 160 μm

with different anode thicknesses was calculated using an electrical circuit simulation tool (PSpice 10.5) under an operating condition of 0.5 V. The equivalent circuit was built up by connecting the sliced circuits with 1 mm thickness in the channel direction, comprising material ohmic resistances and cell (electrochemical) resistance. The parameters used in this calculation are listed in Table 1. These parameters are obtained from our experimental measurements using tubular single cells.

RESULTS AND DISCUSSION
Electrical performance of the prepared cathode-supported honeycomb SOFC

Figure 2 shows the electrical performance of the prepared LSM-supported honeycomb SOFC at 500 and 600 °C. The thickness of anode layer on the channel electrolyte surface was about 10 μm. And the electrode area/ honeycomb volume ratio of the measured honeycomb SOFC was about 20 cm^2/cm^3. The open circuit voltage (OCV) was about 1.0 V within a temperature range from 500 °C to 700 °C. This result indicates that the current and gas leakages through the ScSZ electrolyte are negligible and the prepared electrolyte on the honeycomb channel surface is a dense layer without any cracks and defects. The volumetric power densities at 0.7 V of the honeycomb SOFC were 0.061 and 0.243 W/cm^3 at 500 and 600 °C, respectively. And, the volumetric power densities at 0.5 V were 0.182 and 0.306 W/cm^3 at 500 and 600 °C, respectively. These values of the maximum power density are superior to the 0.2 W/cm^3 at 1000°C resulting from the electrolyte-supported honeycomb SOFC prepared by Wetzko et al. [7]. However, for further improvement of the performance, the current collecting problem and activation of the electrode reaction should be solved as discussed below.

Effect of anode composition on the electrical performance of the cathode-supported SOFC

Figure 3 shows I-V characteristics of the micro-tubular cell with different anode compositions. The *anodes 1 and 2* were composed of 50 wt% NiO- 50wt% GDC and 70 wt%

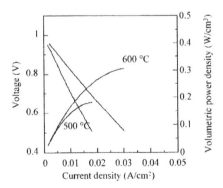

Fig. 2. Performance of a cathode-supported honeycomb SOFC

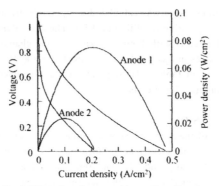

Fig. 3. Effect of anode composition on the cell performance
of cathode-supported single cell

NiO- 30wt% GDC, respectively. The electrical measurement was carried out at 600 °C in
wet hydrogen (3% H_2O) and oxygen atmospheres. The I-V curve of *anode 2* dropped rapidly
at low current densities, mainly due to a lack of the polarization activity and three phase
boundary at the anode side. On the other hand, the application of the high GDC-content
anode (*anode 1*) greatly improves such a rapid I-R drop at low current densities. The power
densities at 0.7 V were 42.8 and 4.7 mW/cm^2 for *anodes 1 and 2*, respectively, which
indicates that *anode 1* can generate more than 9 times higher power density than *anode 2*.
These two cells were composed of the same components, except for the anode layer.
Therefore, the difference in the electrode resistance between two cells was attributed to the
different anode composition. The maximum power densities were 75.2 and 23.8 mW/cm^2 for
anodes 1 and 2, respectively. The maximum power densities from *anode 1* were more than 3
times higher than that from *anode 2*. In the case of the honeycomb SOFC, we select the
composition *anode 2* because of the lack of in-plane conductivity of the composition *anode
1*. For further improvement of the honeycomb SOFC performance, therefore, the current
collecting problem and activation of the electrode reaction at the anode side should be
solved, which is currently being investigated.

Equivalent circuit simulation for cathode-supported honeycomb SOFC

Figure 4 shows the current per honeycomb length for various honeycomb SOFCs
with different honeycomb lengths, which were calculated as a function of the anode
thickness by using PSpice 10.5. The calculation was carried out for the honeycomb SOFCs
with 2×2 square channels under a 0.5 V operation condition, as shown in Fig. 1. Within the
range of few millimeter honeycomb lengths, the current from the honeycomb SOFC was
stable (approximately 0.05A/cm) regardless of the anode thickness. By increasing the
honeycomb length up to 10 mm, however, the current from the honeycomb SOFCs with 10
and 20 μm anode layers largely dropped. This behavior is caused by the lack of in-plane
conductivity of the thin anode layer inside the honeycomb channels. Because the current

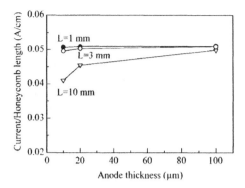

Fig. 4. Calculated current for various cathode-supported honeycomb SOFCs

should pass through the anode film in the in-plane direction, with an increase in the honeycomb length the increase in the current per honeycomb length will be depressed due to the ohmic polarization for passing through the anode film. For further improvement of the honeycomb SOFC performance. we should optimize the channel coating technology in order to increase the anode thickness more than 20 μm (current thickness: 10 μm), which is currently being investigated.

CONCLUSIONS

In this study, we first examined the electrical performances of the cathode-supported honeycomb SOFC with integrated micro-cells. The prepared honeycomb SOFC can generate the volumetric power densities at 0.7 V of 0.061 and 0.243 W/cm^3 at 500 and 600 °C. respectively. The power density of our developed cathode-supported SOFC was affected by the anode composition, and the usage of the high GDC-content anode greatly improves a rapid I-R drop at low current densities. In addition. from the electrical simulations of various honeycomb SOFCs. the anode thickness needed to be more than 20 μm in order to avoid the lack of insufficient in-plane conductivity of thin anode layer.

ACKNOWLEDGEMENT

This study was supported by the New Energy and Industrial Technology Development Organization (NEDO) as part of the Advanced Ceramic Reactor Project.

REFERENCES

[1] I. P. Kibride, Preparation and Properties of Small Diameter Tubular Solid Oxide Fuel Cells for Rapid Start-Up, *J. Power Sources*, **61**, 167-71 (1996).

[2] K. Kendall, and M. Palin, A Small Solid Oxide Fuel Cell Demonstrator for Microelectronic Applications, *J. Power Sources*, **71**, 268-70 (1998).

[3] N. M. Sammes, Y. Du, and R. Bove, Design and Fabrication of a 100W Anode Supported Micro-Tubular SOFC Stack, *J. Power Sources*, **145**, 428-34 (2005).

[4] P. Avila, M. Montes, and E. E. Miró, Monolithic Reactors for Environmental Applications: A Review on Preparation Technologies, *Chem. Eng. J.*, **109**, 11-36 (2005).

[5] S. Shimizu, T. Yamaguchi, T. Suzuki, Y. Fujishiro, and M. Awano, Fabrication and Properties of Honeycomb-Type SOFCs Accumulated with Multi Micro-Cells, *ECS Transactions*, **7**, 661-56 (2007).

[6] T. Yamaguchi, S. Shimizu, T. Suzuki, Y. Fujishiro, and M. Awano, Development of Honeycomb-Type SOFCs with Accumulated Multi Micro-Cells, *ECS Transactions*, **7**, 657-62 (2007).

[7] M. Wetzko, A. Belzner, F. J. Rohr, and F. Harbach, Solid Oxide Fuel Cell Stacks Using Extruded Honeycomb Type Elements, *J. Power Sources*, **83**, 148-55 (1999).

Fabrication

FORMATION OF GAS SEALING AND CURRENT COLLECTING LAYERS FOR HONEYCOMB-TYPE SOFCs

Sota Shimizu[1], Toshiaki Yamaguchi[2], Yoshinobu Fujishiro[2], Masanobu Awano[2]

[1] Fine Ceramics Research Association
Shimoshidami, Moriyama-ku, Nagoya, Aichi, 463-8561 Japan
[2] National Institute of Advanced Industrial Science and Technology (AIST)
2266-98 Anagahora, Shimoshidami, Moriyama-ku, Nagoya, Aichi, 463-8560 Japan

ABSTRACT

In this study, we report the effect of gas sealing layer on the electrical performances of a cathode-supported honeycomb-type SOFC. We have succeeded in preparing a gas sealing layer on the honeycomb edge face of the cathode-supported honeycomb SOFC by developing a stamp process. The prepared honeycomb SOFC was supported by LaSrMnO$_3$ (LSM) honeycomb monolith, and the channel surface was coated with electrolyte/anode bilayer through the use of a new slurry injection method. The electrolyte and anode layers on the channel surface have film thicknesses of about 20 μm and about 10 μm, respectively. The honeycomb SOFC with the gas sealing layer on the edge face exhibited open circuit voltage about 1 V under wet H$_2$ fuel flow, and can generate more than 2.3 times higher power density than the honeycomb SOFC without gas sealing layer.

INTRODUCTION

Recently, SOFCs have received much attention because they are expected to be used as high efficient electrochemical devices, which can convert chemical energy to electrical energy effectively. Among various proposed cell designs, honeycomb-type SOFCs are suitable for compact SOFC modules because of the structural advantages, such as the large capacity of multiple cells and the ease with which the cell size and configuration can be controlled [1]. Thus, the honeycomb-type SOFCs can be beneficial for space saving, thermal control and cost reduction; nevertheless there are few reports on further improvement of the performance of honeycomb-type SOFCs. So far, we reported the technologies to fabricate a LSM-supported honeycomb SOFC via the extrusion of LSM honeycomb monolith and the inner wall coating of multi micro channels using electrolyte and anode slurries [2,3]. The prepared honeycomb SOFC had channel density of about 1000 cpsi (channels per square inch), LSM wall thickness of about 160 μm, a dense ScSZ electrolyte layer about 20 μm, and anode layer about 10 μm. The effective electrode area of the honeycomb (1 cm long, 8×8 mm square, 100 channels) is estimated to be 40 cm^2 per 1 cm^3.

In this study, we investigated the effect of gas leakage through the honeycomb edge face on the electrochemical performances of cathode-supported SOFC using two honeycombs *with and without* a gas sealing layer on the honeycomb edge face.

EXPERIMENTAL PROCEDURE

In this study, $(La_{0.8}Sr_{0.2})_{0.99}MnO_{3\pm\delta}$ (LSM), $Ce_{0.9}Gd_{0.1}O_{1.95}$ (GDC), Sc_2O_3 doped ZrO_2 (ScSZ) and NiO were used as the raw materials. The mixture of LSM powder and cellulose binder was uniaxially extruded through a die to form the LSM honeycomb support. After drying the extruded LSM honeycomb monolith, the channels were coated simultaneously with a ScSZ slurry. As shown in Fig. 1, our proposed cathode-supported honeycomb SOFC also needs a dense gas sealing layer on the edge surface in order to separate air and fuel gases. For this purpose, we introduced a stamp process to prepare a dense ScSZ layer on honeycomb edge face. The LSM honeycomb, which the channels and edge face are coated with ScSZ slurry, was co-fired at 1300 °C for 2 h. For comparison, LSM honeycomb without the edge face coating was also co-sintered at 1300 °C for 2 h in air. Subsequently, the ScSZ-coated LSM honeycombs were coated with a 70 wt% NiO- 30 wt% GDC slurry and heated again at 1300 °C for 2 h. The current-voltage and AC impedance measurements were conducted using a Solartron 1260 frequency response analyzer with a 1296 Interface, using wet H_2 (3 % H_2O) and O_2 as fuel and oxidant, respectively.

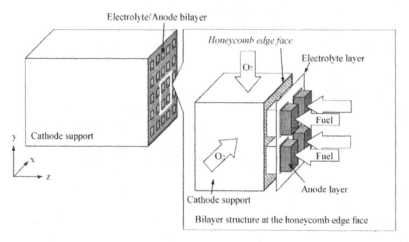

Figure 1. Schematic view of novel SOFC with honeycomb structure, and its enlarged view decomposed to the components

RESULTS AND DISCUSSION

Preparation of gas sealing layer on the edge face of cathode-supported honeycomb SOFC

Figure 2 shows SEM photographs of the cross sections of the prepared honeycomb SOFCs (a) with and (b) without the gas sealing layer on the honeycomb edge face. The electrode-supported honeycomb SOFC has various structural advantages over the electrolyte-supported type. First, the thickness of the electrolyte layer coated on the channel wall can be controlled up to approximately tens of micrometers, while the wall thickness of the electrolyte-supported honeycomb needs to be more than 150 µm to maintain its mechanical strength [4]. Second, the air and fuel gases can be supplied through the channel and honeycomb body. In this case, a fuel gas passes through the channels, then the honeycomb body (wall) can be used as the air supplying pass. Therefore, we should prepare a dense gas sealing layer on the honeycomb edge face to separate air and fuel gases. The edge face of the channel-coated LSM honeycomb using ScSZ slurry was stamped on a ScSZ-ink pad for a few second, and after co-sintering at 1300°C for 2 h, the edge face of the LSM honeycomb wall (160 µm) were successfully coated by a dense gas sealing (ScSZ) layer, as shown in Fig. 2(a). All channels were opened and not clogged by an excess ScSZ coating, because of the usage of a difference of hole size between honeycomb channels ($\approx$ 1.4 mm) and pore in LSM wall (< 10 µm).

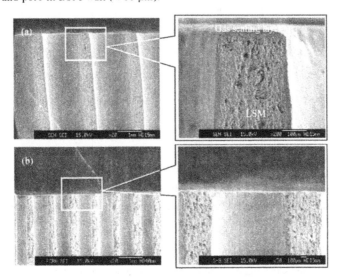

Figure 2. Cross sectional microstructures of the edge area of honeycomb SOFCs. (a) with and (b) without gas sealing layer on the honeycomb edge face

Circuit voltages of the prepared cathode-supported honeycomb SOFCs

Figure 3 summarizes the open circuit voltages (OCVs) of the prepared LSM-supported honeycomb SOFCs (a) with and (b) without the gas sealing layer on the honeycomb edge face in a temperature range between 550 and 650 °C. And the electrode area/ honeycomb volume ratio of the measured honeycomb SOFC was about 25 cm^2/cm^3. The OCV of Fig. 3(a) was about 1.0 V within the measurement temperature range from 550 °C to 650 °C. This result indicates that the current and gas leakages through the ScSZ layer are negligible and the prepared ScSZ on the honeycomb channel surface and edge face is a dense layer without any cracks and defects. In contrast, the honeycomb SOFC without the gas sealing layer exhibited the OCV of 0.77~0.88 V (below 1 V), as shown in Fig. 3(b). The LSM honeycomb wall has a porous structure with 30 vol% open porosity, therefore hydrogen gas can permeate the porous LSM honeycomb body from the honeycomb edge face, which resulted in the low OCV values. The difference between Figs. 3(a) and (b) was decreased with increasing the measurement temperature, which might be due to the increase in the O_2 diffusion through the porous LSM honeycomb body.

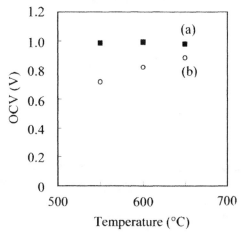

Figure 3. OCV of honeycomb SOFCs (a) with and (b) without the gas sealing layer on the honeycomb edge face

Electrical performances of the prepared cathode-supported honeycomb SOFCs

Figure 4 shows the electrical properties of the honeycomb SOFCs (a) with and (b) without the gas sealing layer. The electrical measurement was carried out at 600 °C in wet hydrogen (3% H_2O) and oxygen atmospheres. I-V curves had almost the same gradient (resistivity) and the curves shifted according to the OCV values. The power densities at 0.7 V were 14 and 6 mW/cm² for the honeycomb SOFCs (a) with and (b) without the gas sealing layer. respectively, which indicates that *honeycomb (a)* can generate more than 2.3 times higher power density than *honeycomb (b)*. From the impedance analysis. ohmic resistance for each cell was almost the same value of about 2.8 Ω·cm². In addition, the electrode polarization impedances of the two honeycomb SOFCs at lower frequency arc also shows the similar behaviors. The power densities at 0.5 V were 17 and 10 W/cm² for the honeycomb SOFCs (a) with and (b) without the gas sealing layer, respectively. For further improvement of the honeycomb SOFC performance, now we are trying to improve the current collecting problem and activation of the electrode reaction at the anode side.

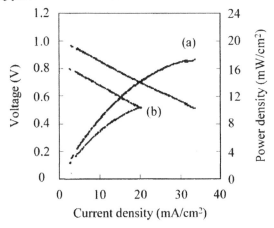

Figure 4. Electrical properties of honeycomb SOFCs (a) with and (b)without the gas sealing layer on the honeycomb edge face

CONCLUSIONS

In this study, we first examined the preparation technology of the gas sealing layer on the edge face of a cathode-supported honeycomb SOFC. Dense ScSZ layer was successfully deposited on the porous honeycomb edge using a stamp process. The cathode-supported honeycomb SOFC with the gas sealing layer is superior to that without the gas sealing layer due to the lack of gas leakage. The honeycomb SOFC with the gas sealing layer on the edge face exhibited open circuit voltage about 1 V under wet H₂ fuel flow, and can generate more than 2.3 times higher power density than the honeycomb SOFC without gas sealing layer.

ACKNOWLEDGEMENT

This study was supported by the New Energy and Industrial Technology Development Organization (NEDO) as part of the Advanced Ceramic Reactor Project.

REFERENCES

[1]P. Avila, M. Montes, and E. E. Miró, Monolithic Reactors for Environmental Applications: A Review on Preparation Technologies, *Chem. Eng. J.*, **109**, 11-36 (2005).

[2]S. Shimizu, T. Yamaguchi, T. Suzuki, Y. Fujishiro, and M. Awano, Fabrication and Properties of Honeycomb-Type SOFCs Accumulated with Multi Micro-Cells, *ECS Transactions*, **7**, 661-56 (2007).

[3]T. Yamaguchi, S. Shimizu, T. Suzuki, Y. Fujishiro, and M. Awano, Development of Honeycomb-Type SOFCs with Accumulated Multi Micro-Cells, *ECS Transactions*, **7**, 657-62 (2007).

[4]M. Wetzko, A. Belzner, F. J. Rohr, and F. Harbach, Solid Oxide Fuel Cell Stacks Using Extruded Honeycomb Type Elements, *J. Power Sources*, **83**, 148-55 (1999).

Characterization
and Testing

Evaluating Redox Stability of Ni-YSZ Supported SOFCs Based on Simple Layer Models

Trine Klemensø[a] and Bent F. Sørensen[b]

Risø National Laboratory – Technical University of Denmark – DTU, 4000 Roskilde, Denmark
[a]Fuel Cells and Solid State Chemistry Department
[b]Materials Research Department

Abstract

The prevalent solid oxide fuel cell design contains two Ni cermet structures: an anode support layer and an active anode layer. The cermets are not redox stable, and bulk expansion occurs when they are subjected to reoxidation. The expansions result in cracking of the critical electrolyte layer.

The stress state of the electrolyte was evaluated based on a fracture energy model and a simple layer model considering the biaxial stresses in the anode support, the active anode and the electrolyte. The stress generated in the electrolyte upon reoxidation depends on the cell design, operational conditions and bulk expansion of the cermets, which in turn is determined by the microstructure and the cermet strength.

A parameter study to investigate how to avoid cracks in the electrolyte was carried out. The technological adjustable parameters, i.e. the operational temperature, the thickness of anode support and the cermet bulk expansions, were chosen and varied. According to the model, lower temperatures and thinner anode supports enables the avoidance of cracking. At 650°C the model predicted that cracking could be avoided with minor improvements in the bulk expansions. If the bulk expansion of the anode support was reduced from the present 0.35% to 0.2%, and the anode bulk expansion was reduced from 1.19% to below 0.7%, cracks in the electrolyte will be prevented and the cell redox stable according to the model.

Introduction

The state-of-the-art solid oxide fuel cell (SOFC) consists of a Ni-YSZ (yttria stabilized zirconia) anode support, an active Ni-YSZ anode, an YSZ electrolyte, and a LSM (lanthanum strontium manganite)-YSZ cathode [1]. Redox stability of the anode parts are considered to be a commercial requirement for many applications of the SOFC technology [2,3]. However, oxidation of the Ni-YSZ cermets in the anode supported design are detrimental to the cell performance [4,5,6]. The degradation is related to a bulk expansion of the cermet structures upon the re-oxidation, which generates critical cracks in the electrolyte layer [4,7,8]. Figure 1 shows micrographs of a cell subjected to nickel reoxidation. Layer #1 is the electrolyte, #2 is the active anode and #3 is the anode support. The cracks were observed to be restricted to the thin electrolyte layer.

It has been shown that the bulk expansions of the Ni-YSZ cermets are related to the increase in volume when the Ni particles are reoxidized to NiO in combination with morphological changes of the nickel particles during redox cycling. The mechanisms are more thoroughly described in [9,10]. As a result, the cermet layers experience a volume expansion that can be expressed as an inelastic (dilatation) strain.

The purpose of the present work was to evaluate the tolerated cermet bulk expansions based on a simple layer model. The model considered the biaxial stresses in three of the layers: anode support, active anode, and electrolyte. In addition, the problem was evaluated based on a fracture energy model. The problem was investigated for variable cell designs (i.e. anode support thickness) and different operational temperatures.

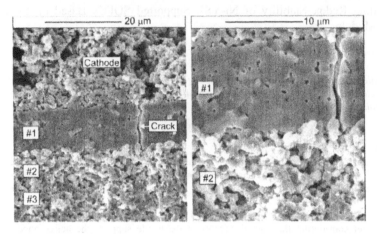

Figure 1. SEM micrographs of a full cell i.e. including the cathode layer after reoxidation. The electrolyte is denoted layer #1, layers #2 is the active anode, layer #3 is the anode support.

Problem definition

The cell was modeled as a layered system, where the single layers represent the anode support, the anode, and the electrolyte, respectively. The cathode was not included; since it is thin and possesses a low stiffness, the influence of the layer on the stress state of the cell is believed to be insignificant [11]. The system is sketch in Figure 2a for a unit cell. The unit cell is representative for the behavior of the SOFC cell remote from the edges when the cell is kept flat. The layers are stress-free at the processing temperature at 1300°C (cf. Figure 2a). In the following, each layer will be treated as homogeneous materials having isotropic elastic properties. Out of plane strain changes are "free" in the sense that they do not induce constraints, and therefore do not affect the resulting stresses.

Inelastic strain is induced when the temperature is lowered to the operational temperature around 850°C, and possibly when the NiO-YSZ cermets are reduced. The situation is illustrated in Figure 2b for the case where the layers are free, i.e. not coherent. The thermal expansion coefficient (TEC) of the electrolyte is lower than for the cermets, both in the sintered and reduced state. The reduction involves shrinkage of the nickel phase, which also may contribute to the strain.

Re-oxidation of the cermets will be associated with inelastic strains. The bulk of the cermets will expand, whereas the electrolyte will remain unaffected. This is shown in Figure 2c, where the layers are free to expand.

The layers in the cell are co-sintered at the processing temperature, and will remain coherent during the different operational steps. Thus, the layers will experience a common strain (ε_0) upon operation as illustrated in Figure 2d. The TEC mismatch, the reduction, and the mounting of the cell into a stack, which involved flattening by mechanical load and restrictions by other stack components, will contribute with compressive stress to the electrolyte, whereas the re-oxidation involves a tensile stress contribution to the electrolyte.

The layers can be assumed to remain planar during the processes. This is reasonable, since the cell is fixed to planarity when mounted in a SOFC stack. The layers were also assumed not to be restricted by other components in the stack. Thus, the dimensional change of the layers is solely determined by the mutual restrictions between the layers in the system.

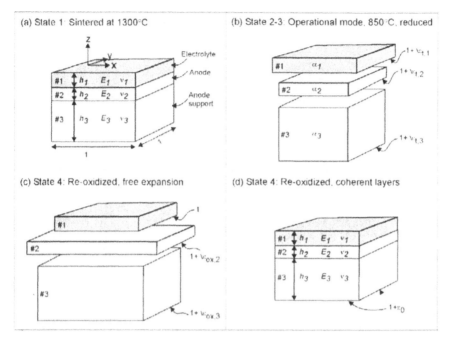

Figure 2. Sketch of the system and the history it experiences. (a) State 1 is the stress-free system at the sintering temperature around 1300°C. Layer 1 represents the electrolyte, layer 2 the active anode, and layer 3 the anode support. Each layer is assigned a thickness (h_i), an elastic modulus (E_i), and a Poisson's ratio (v_i). The lateral dimension of the as-sintered layers is of unit length, and only the lateral dimensional changes are indicated in the following figures. (b) State 2-3 is the operational mode, where the temperature has been reduced to 850°C (state 2), and the nickel phase in layers 2 and 3 has been reduced to metallic Ni (state 3). The state is shown with free contraction of the layers. The thermal strain is denoted $\Delta\varepsilon_{t,i}$. State 4 is the re-oxidized state, illustrated respectively when the layers are free to expand (c), and when the layers are coherent (d).

The inelastic strain under free expansion, $\Delta\varepsilon_i$, is the sum of the thermal induced strain, and the strain related to the phase-transition, i.e. the nickel oxidation. Since the lateral dimensions are big compared to the layer thicknesses, the effect of the stress-free edges is insignificant, and it is valid to consider only the linear dimensional changes. Thus, the strain can be expressed as in Equation 1, where $\Delta\varepsilon_{t,i}$ represents the thermal induced strain, and $\Delta\varepsilon_{ox,i}$ is the strain introduced upon re-oxidation. α is the thermal expansion coefficient, ΔT is the difference between the actual temperature (T), and the temperature for the stress-free state (T_0).

(Eq. 1) $\qquad \Delta\varepsilon_i = \Delta\varepsilon_{t,i} + \Delta\varepsilon_{ox,i} = \alpha_i \cdot \Delta T + \Delta\varepsilon_{ox,i}$

Model solution

Each layer may be assumed to be isotropic and plane, and thus, biaxial stress occurs away from the boundaries. The biaxial stress state is described by Equation 2, where σ denotes stress, and ε denotes strain. The subscripts xx and yy refer to the normal stress or strain in the x-direction and y-direction, respectively.

(Eq. 2) $\qquad \sigma_{xx}(z) = \sigma_{yy}(z) \quad , \qquad \varepsilon_{xx} = \varepsilon_{yy}$

The relationship between the biaxial stress and strain is described by Hooke's law, which is shown in Equation 3.

(Eq. 3) $\varepsilon_{xx} = \dfrac{\sigma_{xx}}{E} - \dfrac{v \cdot \sigma_{yy}}{E} + \Delta\varepsilon$

E is the elastic modulus, v is the Poisson's ratio, and $\Delta\varepsilon$ represents the inelastic strain under free expansion (cf. Figures 2b and 2c).

The stresses were assumed constant, and independent of the z-position in each layer. Thus, inserting Equation 2 into Equation 3 leads to:

(Eq. 4) $\varepsilon_1 = \dfrac{\sigma_1}{E_1} \cdot (1 - v_1) + \Delta\varepsilon_1$

For the case where the three layers are coherent, illustrated in Figure 2d, the strain must be identical for each layer (denoted ε_0), as shown in Equation 5:

(Eq. 5) $\varepsilon_1 = \varepsilon_2 = \varepsilon_3 = \varepsilon_0$

By inserting Equation 5 into Equation 4 and rearranging, an expression for the stress in each layer is obtained. Equation 6 shows the obtained result for the stress in layer 1. For convenience, the abbreviation E_i^* as defined in Equation 7, is used. This is possible under the restriction that v_i is not equal to unity. Subscript i denotes layer number.

(Eq. 6) $\sigma_1 = \varepsilon_0 \cdot \dfrac{E_1}{1 - v_1} - \dfrac{E_1}{1 - v_1} \cdot \Delta\varepsilon_1 = \varepsilon_0 \cdot E_1^* - E_1^* \cdot \Delta\varepsilon_1$

(Eq. 7) $E_i^* = \dfrac{E_i}{1 - v_i}$

Since planarity and no restrictions from the surroundings were assumed, the forces in the layers will outbalance each other:

(Eq. 8) $\sigma_1 \cdot h_1 + \sigma_2 \cdot h_2 + \sigma_3 \cdot h_3 = 0$

Equation 6 and the analogous expressions for the stress in layer 2 and 3 are inserted into Equation 8:

(Eq. 9) $\varepsilon_0 = \dfrac{E_1^* \cdot \Delta\varepsilon_1 \cdot h_1 + E_2^* \cdot \Delta\varepsilon_2 \cdot h_2 + E_3^* \cdot \Delta\varepsilon_3 \cdot h_3}{E_1^* \cdot h_1 + E_2^* \cdot h_2 + E_3^* \cdot h_3}$

The obtained expression for ε_0 is substituted into Equation 6. From this, an expression for the stress in layer 1 (the electrolyte), as a function of the free expansion, layer thickness, and elastic properties of the layers is obtained (Equation 10).

(Eq. 10) $\sigma_1 = \dfrac{E_1^* \cdot E_2^* \cdot h_2 \cdot (\Delta\varepsilon_2 - \Delta\varepsilon_1) + E_1^* \cdot E_3^* \cdot h_3 \cdot (\Delta\varepsilon_3 - \Delta\varepsilon_1)}{E_1^* \cdot h_1 + E_2^* \cdot h_2 + E_3^* \cdot h_3}$

When combining Equations 1 and 10, the stress in the electrolyte can be expressed:

(Eq. 11)

$$\sigma_1 = \frac{E_1^* \cdot E_2^* \cdot h_2 \cdot \left(\Delta T \cdot (\alpha_2 - \alpha_1) + \Delta\varepsilon_{ox,2} - \Delta\varepsilon_{ox,1}\right) + E_1^* \cdot E_3^* \cdot h_3 \cdot \left(\Delta T \cdot (\alpha_3 - \alpha_1) + \Delta\varepsilon_{ox,3} - \Delta\varepsilon_{ox,1}\right)}{E_1^* \cdot h_1 + E_2^* \cdot h_2 + E_3^* \cdot h_3}$$

Energy balance model

When considering fracture, two conditions must be met before cracking occurs. First, a crack must form from the microstructure, e.g. origination from a pore or a flaw. Usually, the initiation of crack growth is described in terms of a critical stress criterion. Second, once a sharp crack has formed, it must be able to propagate. The criterion for propagation of a crack is that the decrease in potential energy is equal to the energy consumed. The potential energy released is called the energy release rate (G). Thus, for the electrolyte to fracture, the stress must exceed both the critical stress for crack initiation and the stress for crack propagation.

Generally G depends on the stress level as well as the sample geometry and crack size. The energy expended to create cracking is a material property. The property is described by the critical energy release rate (G_{IC}) that is the energy used per crack area formed, and has the unit J/m^2. For ceramics the energy is primarily consumed to form new surface. Depending on the material, G_{IC} can be constant or increase with the size of the crack. Instable, or steady state cracking will occur independent of the crack size when $G \geq G_{IC}$.

For a 2-layered system, consisting of a film (layer 1) of thickness, h, on a semi-infinite substrate (corresponding to the combined layers 2 and 3 in Figure 2a), the energy release rate attains a steady state value as the crack length is longer than a few times the layer thickness. The steady state energy release rate of a crack in the film was calculated by Beuth (cf. Equation 12) [12].

(Eq. 12) $$G = \frac{1}{2} \cdot \frac{\sigma_1^2 \cdot h_1}{\overline{E_1}} \cdot \pi \cdot g(\alpha, \beta)$$

In the equation σ denotes the tensile stress prior to the introduction of the crack, $\overline{E_1}$ is defined in Equation 13, and $g(\alpha, \beta)$ was tabulated by Beuth [12]. The so called Dundur's parameters [13] α and β are defined in Equations 14 and 15. In analogy with the previous section, E_i is the elastic modulus of layer i, v is Poisson's ratio, μ is the shear modulus of the material, which can be calculated according to Equation 16, and the subscript 2 refers to the substrate.

(Eq. 13) $$\overline{E_1} = \frac{E_1}{1 - v^2}$$

(Eq. 14) $$\alpha = \frac{\overline{E_1} - \overline{E_2}}{\overline{E_1} + \overline{E_2}}$$

(Eq. 15) $$\beta = \frac{\mu_1 \cdot (1 - 2 \cdot v_2) - \mu_2 \cdot (1 - 2 \cdot v_1)}{2 \cdot \mu_1 \cdot (1 - v_2) + 2 \cdot \mu_2 \cdot (1 - v_1)}$$

(Eq. 16) $$\mu_i = \frac{E_i}{2 \cdot (1 + v_i)}$$

It should be noted that the model by Beuth is only valid for cracks restricted to the thin layer, where the crack tip is at the interface. If the crack is deflected at the interface or propagates into the substrate, the energy release rate can be higher, and this must be taken into account as reported by Ye et al.[14].

Application of the models

To apply the models, it is necessary to know the characteristic material properties. The properties will depend on the composition and porosity of the materials. Data on materials as close as possible to the prevalent state-of-the-art materials were used, and are summarized in Table 1.

Thermal expansion coefficients for the separate layers 2 and 3 were not found, and they were assigned identical values ($\alpha_{anode,ox}$).

The fracture strength of the electrolyte ($\sigma_{f,1}$) as reported in [19] was measured by ring-on-ring measurements on 150 µm thick samples. The radii of the two rings were respectively 1.7 mm and 8.5 mm. However, the reference volume was not reported. Considering the volume within the loading ring, the test volume is much bigger than the average electrolyte that has a thickness in the range of 10 µm and a radius in the same range.

The model by Beuth is a valid model for the redox problem with SOFCs since the cracks in the electrolyte are restricted to the thin electrolyte layer (cf. Figure 1). The parameters α and β were calculated to 0.47 and 0.22, respectively, based on Equations 14 and 15, and Table 1. The function $g(\alpha,\beta)$ was approximated to 1.949 as the closest tabulated values to the calculated α and β values were 0.50 and 0, respectively [9]. The shear moduli μ_1, μ_2, and μ_3 were calculated using Equation 16 and the values in Table 1, to respectively 59 GPa, 46 GPa, and 26 GPa.

Table 1. Material properties for the layers in the models. E is Young's modulus, v is Poisson's ratio, h is layer thickness, $\Delta\varepsilon_{ox,2}$ and $\Delta\varepsilon_{ox,3}$ are the strain related to the reoxidation of layer 2 and 3, $\alpha_{anode,ox}$ and $\alpha_{anode,red}$ are the thermal expansion coefficient of the anode in the oxidized and reduced state respectively, $\sigma_{f,1}$ is the fracture strength of the electrolyte, and G_{IC} is the critical energy release rate for the electrolyte. The subscript refers to the number of the layer, and RT signifies room temperature.

Property	Value	Material	Test conditions	Ref.
E_1 [GPa]	155	8YSZ, dense	900°C, impulse excitation technique	[19]
E_2 [GPa]	120	NiO-8YSZ, 12% porosity	RT, estimated	[16]
E_3 [GPa]	60	NiO-3YSZ, 17% porosity	RT, strain gauge	[16]
v_1	0.315	8YSZ, dense	-	[19]
v_2	0.317	NiO-8YSZ, 11% porosity	-	[20]
v_3	0.17	NiO-8YSZ, 31% porosity	-	[20]
h_1 [µm]	10	-	-	[21]
h_2 [µm]	10	-	-	[21]
h_3 [µm]	300	-	-	[21]
$\Delta\varepsilon_{ox,2}$ [%]	1.19	NiO-8YSZ, 2% porosity	1000°C, dilatometry	[8]
$\Delta\varepsilon_{ox,3}$ [%]	0.35	NiO-3YSZ, 8% porosity	1000°C, dilatometry	[8]
α_1 [K^{-1}]	$10.9\cdot10^{-6}$	YSZ	-	[22]
$\alpha_{anode,ox}$ [K^{-1}]	$12.6\cdot10^{-6}$	NiO-YSZ	-	[22]
$\alpha_{anode,red}$ [K^{-1}]	$13.2\cdot10^{-6}$	Ni-YSZ	-	[22]
$\sigma_{f,1}$ [MPa]	265	8YSZ, dense	900°C, ring-on-ring	[19]
G_{IC} [J/m^2]	2.8	8YSZ, <2% porosity	RT, DCB (double cantilever beam)	[21]

For the cell to be operational and redox stable, it is critical that cracks in the electrolyte do not form. Thus, the stress in the electrolyte must remain below the fracture strength, or below the critical stress (corresponding to the critical layer thickness), during the history of the cell.

Figure 3 (left) illustrates the stress in the three layers (σ_i) as a function of state as predicted from Equation 11. The corresponding stress levels in the layers 2 and 3 were calculated analogous to Equation 11, and the values were checked according to Equation 8. Fractures in these layers are less critical as porosity is required for the electrochemical reaction to occur. The fracture strength of the electrolyte, and the critical stress according to Equation 12, are included in Figure 3. The Figure to the right illustrates the strains when the layers are not restricted by each other.

In the sintered state, at the processing temperature around 1300°C, the specimen is stress-free (state 1). The cell is operated at a much lower temperatures, and the nickel in the anode cermets is to be reduced before operation (state 2-3). When the temperature is reduced to the operation mode 850°C, compressive stress arises in the electrolyte (-158 MPa), and tensile stresses in the two other layers (12 MPa in layer 2, 5 MPa in layer 3) due to the TEC mismatch (cf. Table 1). The reduction of the nickel part of the cermet may also generate stresses, since it involves changes in the volume, and shrinkage of the nickel network. However, the contribution from the reduction is believed to be minor. For instance, no change in the stress state upon reduction has been reported by Yakabe et al. [15,16], and a tensile stress in the range of 40 MPa upon reduction was reported by Fischer et al. [17]. In Figure 3, the effect of the reduction was assumed to be insignificant.

In the case of shut-down of the system, the cermet components may become re-oxidized (state 4). The re-oxidation of the nickel involves bulk expansion of the cermet structures, whereas the electrolyte remains unaffected (i.e. $\Delta\varepsilon_{ox,1} = 0$). For cermets approximating the anode support and the active anode the strain has been reported to be 0.35% and 1.19% upon oxidation (cf. Table 1). However, the strain is known to depend on the microstructure, and factors such as particle size distributions, local porosity, ceramic composition, and temperature [2,4,8,18]. The bulk expansions generate tensile stress in the electrolyte, and from Figure 3 it is seen that the stress in layer 1 in state 4 exceeds both the critical stress and the fracture stress of the electrolyte. Thus, the model predicts cracking of the electrolyte. The model results agree with experimental studies (cf. Figure 1), and the reoxidation can therefore explain the observed fractures, and the lack of redox stability of the system.

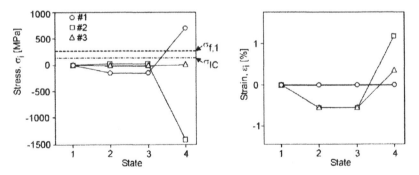

Figure 3. Left: the stress state of the three layers during the history of a state-of-the-art anode-supported cell. The fracture strength ($\sigma_{f,1}$), and the critical stress level (σ_{IC}) of layer 1 are included. State 1 is the as-sintered state at the processing temperature of 1300°C. In state 2 the temperature has been reduced to 850°C. In state 3 the cermets have been activated. State 4 is the re-oxidized state. Right: illustration of the strains when the layers are unrestricted.

Parameter study

Redox stability may be achieved by adjusting some of the operational parameters or modifying the materials. The technological adjustable parameters are believed to be restricted to the operational temperature (T), the thickness of the anode support (h_3), and the strain related to the reoxidation ($\Delta\varepsilon_{ox.2}$ and $\Delta\varepsilon_{ox.3}$). The range for variation in the parameters will also be limited. For the operation, temperatures in the range 650°C-1000°C are considered realistic. Regarding the thickness of the anode support, it may either be absent (corresponding to a metal-supported cell design), or up to 1 mm in thickness. For the bulk expansion, microstructural optimizations where the expansion is eliminated, i.e. $\Delta\varepsilon_{ox.2}$ and $\Delta\varepsilon_{ox.2}$ approaching 0, are believed to be possible.

Based on this, a parameter study on the dependence of the stress in the electrolyte as a function of the design, operation, and material properties, was carried out. The study was restricted to nine basic scenarios based on three different operational temperature ($T = 650°C$, 850°C, and 1000°C), and three anode support thickness ($h_3 = 0$ μm, 300 μm, and 1000 μm). Within each scenario, the bulk expansion of layer 2 ($\Delta\varepsilon_{ox.2}$) was varied from 0 to 1.19%, and the bulk expansion of layer 3 ($\Delta\varepsilon_{ox.3}$) was varied between 0 and 0.35%.

$h_3 = 0$ μm

Figure 4 shows the case, where layer 3 has been eliminated, i.e. $h_3 = 0$ μm. The stress states of layer 1 as a function of the bulk expansion of layer 2, for the three different temperatures, are shown. The fracture strength of the electrolyte ($\sigma_{f.1}$) and the critical stress of the electrolyte (σ_{IC}) are indicated in the Figure.

For the system without an anode support, the tolerated bulk expansion of the active anode ($\Delta\varepsilon_{ox.2}$) is much lower than the state-of-the-art value of 0.0119 (cf. Table 1), irrespective of the operating temperature. The bulk expansion should be reduced to around 0.2-0.35% to avoid fractures upon reoxidation according to the two models.

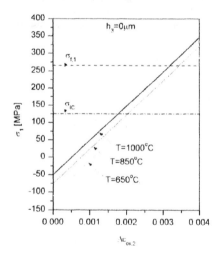

Figure 4. The stress state of layer 1, when layer 3 omitted ($h_3 = 0$ μm), as a function of the expansion in layer 2, shown for the three operational temperatures. The fracture strength of the electrolyte and the critical stress of the electrolyte are indicated.

$h_3 = 300$ µm

If the thickness of layer 3 is chosen to be 300 µm as in the state-of-the-art design, the development in the stress state of layer 1 as a function of the expansions of layer 2 and 3, is shown in Figures 5a-c for the three different temperatures: (a) is at 650°C, (b) is at 850°C, and (c) is at 1000°C. The cases where $\Delta\varepsilon_{ox,3}$ is 0, 0.001, 0.002, and 0.003 are shown, and the fracture strength and critical stress for the electrolyte are included.

At 650°C the electrolyte will be stressed below the fracture strength if the two bulk expansions can be reduced, $\Delta\varepsilon_{ox,3}$ to 0.002 and $\Delta\varepsilon_{ox,2}$ to < 0.007. Further, if $\Delta\varepsilon_{ox,3}$ can be reduced to 0.001, the stress will be below the critical stress (cf. Figure 5a).

(a)

(b)

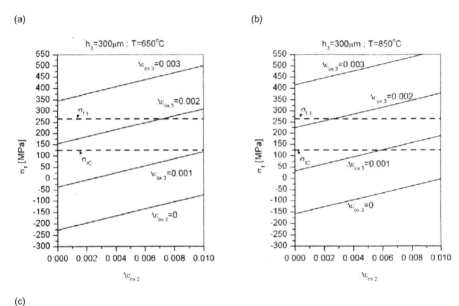

(c)

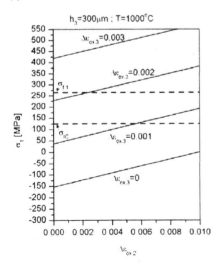

Figure 5. The stress state of layer 1 as a function of the expansion in layer 2 and 3, when the thickness of layer 3 is set at 300 µm. The situation is shown for three temperatures: (a) 650°C, (b) 850°C, and (c) 1000°C. The fracture strength of the electrolyte and the critical stress of the electrolyte are indicated.

At the higher temperatures the required improvements to obtain redox stability are also increased. To be below the fracture strength at 850°C and with $\Delta\varepsilon_{ox,3}$ reduced to 0.002, requires $\Delta\varepsilon_{ox,2}$ to < 0.003 (cf. Figure 5b).

$h_3 = 1000 \, \mu m$

If the thickness of layer 3 is increased to 1000 μm, the development in σ_1 as a function of the expansions of layer 2 and 3, for the three different temperatures, are shown in Figures 6a-c.

(a)

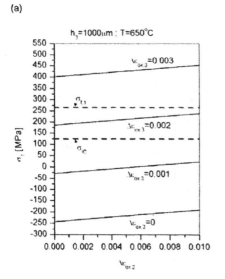

(b)

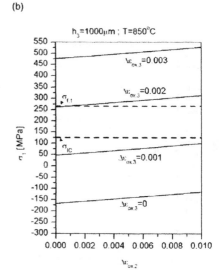

(c)

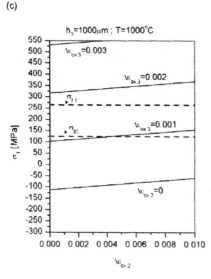

Figure 6. The stress state of layer 1 as a function of the expansion in layer 2 and 3, when the thickness of layer 3 is set at 1000 μm. The situation is shown for three temperatures: (a) 650°C, (b) 850°C, and (c) 1000°C. The fracture strength of the electrolyte and the critical stress of the electrolyte are indicated.

The cases where $\Delta\varepsilon_{ox.3}$ is 0, 0.001, 0.002, and 0.003 are shown, and the fracture strength and critical stress for the electrolyte are included.

At 650°C the stress will be below the fracture stress if $\Delta\varepsilon_{ox.3}$ is reduced to 0.002. For the electrolyte to be below the critical stress, then $\Delta\varepsilon_{ox.3}$ must be further reduced to between 0.001-0.002 (cf. Figure 6a).

Again, higher temperatures tighten up the requirements. At 850°C and with $\Delta\varepsilon_{ox.3} = 0.002$, then $\Delta\varepsilon_{ox.2}$ must be < 0.001 (cf. Figure 6b), and at 1000°C it is not possible to be below the fracture strength with $\Delta\varepsilon_{ox.3} = 0.002$ (cf. Figure 6c).

Conclusions

From the models and the parameter study it was possible to see the effect of different technological parameters on the redox stability.

When excluding the anode support (layer 3) and disregarding another substrate material, the required improvement of the bulk expansion of the anode was quite big. The bulk expansion should be reduced from the present 1.19% to at least 0.3%.

For the designs including the anode support, the required improvements in the bulk expansions were bigger the higher temperature and the thicker the anode support. However, at 650°C the required improvements do not seem out of reach. The models indicated that if the bulk expansion of the anode support can be reduced from the present 0.35% to 0.2%, and the anode bulk expansion is reduced to below 0.7%, redox stability is achieved.

The models do not include the stresses arising during mounting of the cell in the stack, weight and other restrictions from the external stack components. Further, only one redox cycle was considered.

References

1. S. C. Singhal, K. Kendall, High temperature solid oxide fuel cells, Elsevier Ltd., Oxford, UK (2003).
2. G. Robert, A. Kaiser, E. Batawi, Anode substrate design for RedOx-stable ASE cells, in 6th European SOFC Forum proc. vol. 1, M. Mogensen (eds.), European Fuel Cell Forum, Oberrohrdorf, Switzerland, 193-200 (2004).
3. S. Tao, J. T. S. Irvine, A redox-stable efficient anode for solid-oxide fuel cells, Nature Materials, 2, 320-323 (2003).
4. D. Fouquet, A. C. Müller, A. Weber, E. Ivers-Tiffée, Kinetics of oxidation and reduction of Ni/YSZ cermets, Ionics, 9, 103-108 (2003).
5. M. Cassidy, G. Lindsay, K. Kendall, The reduction of nickel-zirconia cermet anodes and the effects on supported thin electrolytes, Journal of Power Sources, 61, 189-192 (1996).
6. D. Waldbillig, A. Wood, D. G. Ivey, Electrochemical and microstructural characterization of the redox tolerance of solid oxide fuel cell anodes, Journal of Power Sources, 145, 206-215 (2005).
7. G. Stathis, D. Simwonis, F. Tietz, A. Moropoulou, A. Naoumides, Oxidation and resulting mechanical properties of Ni/8Y2O3-stabilized zirconia anode substrate for solid-oxide fuel cells, Journal of Materials Research, 17, 951-958 (2002).
8. T. Klemensø, C. Chung, P. H. Larsen, M. Mogensen, The mechanism behind redox instability of anodes in high-temperature SOFCs, Journal of the Electrochemical Society, 152, A2186-A2192 (2005).
9. T. Klemensø, C. C. Appel, M. Mogensen, In situ observations of microstructural changes in SOFC anodes during redox cycling, Electrochemical and Solid-State Letters, 9 (9), A403-A407 (2006).

10. T. Klemensø, M. Mogensen, Ni-YSZ solid oxide fuel cell anode behavior upon redox cycling based on electrical characterization, Journal of the American Ceramic Society, 90 (11), 35-82-3588 (2007).

11. R. Martins, A. Hagen, V. Honkimäki, H. F. Poulsen, R. Feidenhans'l, In-situ study of residual strain in solid oxide fuel cells (2005).

12. J. L. Beuth Jr., Cracking of thin bonded films in residual tension, International Journal of Solids Structures, 29, 1657-1675 (1992).

13. J. Dundurs, Edge-bonded dissimilar orthogonal elastic wedges under normal and shear loading, Journal of Applied Mechanics, 36, 650-652 (1969).

14. T. Ye, Z. Suo, A. G. Evans, Thin film cracking and the roles of substrate and interface, International Journal of Solids Structures, 29, 2639-2648 (1992).

15. H. Yakabe, Y. Baba, T. Sakurai, M. Satoh, I. Hirosawa, Y. Yoda, Evaluation of residual stresses in a SOFC stack, Journal of Power Sources, 131, 278-284 (2004).

16. H. Yakabe, Y. Baba, T. Sakurai, Y. Yoshitaka, Evaluation of the residual stress for anode-supported SOFCs, Journal of Power Sources, 135, 9-16 (2004).

17. W. Fischer, J. Malzbender, G. Blass, R. W. Steinbrech, Residual stresses in planar solid oxide fuel cells, Journal of Power Sources, 150, 73-77 (2005).

18. D. Waldbillig, A. Wood, D. G. Ivey, Thermal analysis of the cyclic reduction and oxidation behaviour of SOFC anodes, Solid State Ionics, 176, 847-859 (2005).

19. A. Selcuk, A. Atkinson, Strength and toughness of tape-cast yttria-stabilized zirconia, Journal of the American Ceramic Society, 83, 2029-2035 (2000).

20. N. M. Sammes, Y. Du, The mechanical properties of tubular solid oxide fuel cells, Journal of Materials Science, 38, 4811-4816 (2003).

21. A. N. Kumar, B. F. Sørensen, Fracture energy and crack growth in surface treated yttria stabilized zirconia for SOFC applications, Materials Science and Engineering, A333, 380-389 (2002).

22. J. Malzbender, R.W. Steinbrech, L. Singheiser, Journal of Material Research ,18 (4), 929-934 (2003).

DEGRADATION PHENOMENA IN SOFCS WITH METALLIC INTERCONNECTS

Norbert H. Menzler, Frank Tietz, Martin Bram, Izaak C. Vinke, L.G.J. (Bert) de Haart

Forschungszentrum Jülich, Institute of Energy Research, Jülich, GERMANY

ABSTRACT

The long-term operation of SOFCs involves various degradation phenomena. They can be divided into two major groups: one is internal phenomena which are related to interactions between the layers or microstructural or chemical changes during operation. The other is external parameters such as the chosen operational conditions (e.g. applied current, temperature, time) or foreign elements in the fuel gas or in the air (e.g. carbon, sulphur, chromium).

This presentation gives an overview of possible causes of degradation when using metallic interconnects and focuses on two phenomena; one is the incorporation of volatile Cr species into the cathode and the other is the microstructural change of the cathodic contact layers.

For chromium poisoning it was found that the degradation of stacks is not directly related to the amount of Cr incorporated in the cathode but mainly depends on the current density applied. The higher the current density, the higher the degradation rate or the increase of the area-specific resistance becomes.

With respect to cathodic contacting it should be noted that protection layers based on powders tend to agglomerate during stack operation. The direct influence of this reduction in contact area on stack degradation is not yet measurable.

The contact layers themselves remain microstructurally unchanged but Cr evaporating from the interconnect is incorporated into them.

INTRODUCTION

Up to now, the market entry of solid oxide fuel cell (SOFC) systems has been hampered by the degradation of cells and stacks. Especially for stationary applications, more than 40,000h of operation under nearly steady-state conditions are envisaged. For auxiliary power unit (APU) applications, shorter operating time is planned (5,000-8,000h), but in this case thermal and redox cyclability has to be taken into account.

Degradation rates of state-of-the-art SOFC stacks with metallic interconnects are still about 1% per 1,000h (in terms of voltage degradation), which is hardly tolerable for stationary applications. Degradation rates normally increase when thermal or redox cycling is applied. Degradation rates of about 3-5% per 1,000h are thus not acceptable for APU application.

An overview is presented of the parameters which affect degradation of cells and stacks, highlighting novel scientific results with respect to degradation phenomena. Additionally, a critical examination of the measurement methods and data analysis is given.

Two examples are presented in more detail: cathode poisoning by chromium species on the stack level and the degradation of interconnect coatings with respect to microstructural evolution.

At Forschungszentrum Jülich stacks with cell sizes of 100x100mm² and 200x200mm² are routinely operated. The stack size with the smaller cells represents the R&D stacks for characterizing the influence of single parameters during operation (e.g. temperature, time, fuel gas, fuel or air utilization...) or the use of novel stack components (cells, sealant, interconnect, contacts, materials) or component microstructures, thicknesses etc. Stacks operated with the bigger cells are used for demonstration purposes, which include stacking of large plates and large sizes (repeating units) and operation under simulated system conditions although still in a furnace. Since the cell, sealant and interconnect materials show good R&D performance and stack assembling and testing reveals good reproducibility, long-term operational experience under varying operational conditions and the influence on stack behaviour (degradation, failure, and running-in times) is attracting ever greater interest.

Table 1: Possible reasons for degradation of SOFC stacks during operation

Cell or layer inherent causes		
SOFC part	Change	Possible causes
Cathode	Change in chemical composition of cathode	Interdiffusion, e.g. Mn depletion for LSM, Sr depletion for LSFC or Gd interdiffusion for CGO/LSFC layers
	Decomposition	Formation of La oxide, formation of Co oxide
	Coarsening	Post-sintering effect due to operating temperature leading to reduction of triple-phase boundaries (TPBs)
	Interactions	With electrolyte material, barrier layer (for LSFC) or cathode contact layer
Electrolyte	Change of crystal structure	Crystal instabilities (cubic phase) due to phase destabilization (change to tetragonal/monoclinic) because of inhomogeneities of the stabilizing agent or insufficient amount of stabilization agent
	Incorporation of foreign elements	Diffusion of cathode or anode elements at the grain boundaries or dissolution into the YSZ (influence on ionic conductivity; appearance of electronic conduction leading to cell-inherent short circuiting)
	Segregations	Segregation of impurities (e.g. SiO2) at the grain boundaries
Anode	Coarsening	Post-sintering effect due to operating temperature leading to reduction of triple-phase boundaries (TPB)
	Ni evaporation	Reduction of electronic conductivity, loss of percolation, reduction of TPBs
Substrate	Ni evaporation	Reduction of electronic conductivity, loss of percolation, reduction of reforming activity
	Coarsening	Post-sintering effect due to operating temperature
Anodic contacting	Formation of less-conducting oxide phases	Corrosion on interconnect or between interconnect and e.g. nickel mesh
	Contact loss	Due to thermal cycles (differences in coefficient of thermal expansion, CTE) or creep

Table 1 cont.: Possible reasons for degradation of SOFC stacks during operation

SOFC part	Change	Possible causes
Cathodic contacting	Interactions	Between oxide phases on interconnect, evaporation protection layers (PL), contact layers (CCL) and current collector/cathode
	Mechanical contact loss	Due to shrinking caused by post-sintering effects, phase reactions, interdiffusions, Ostwald ripening, evaporation, creep
	Increase of resistance	Contact layer inherent ageing, corrosion of interconnect. Chemical reactions between layers.
External causes		
Cathode	Incorporation of foreign elements	Chromium poisoning may lead to reduction of TPBs, formation of less-conducting phases, reduction of open porosity (diffusion hindrance), decomposition
	Incorporation of foreign elements	Coarsening due to incorporation of elements into the lattice (change of CTE, sinterability...)
	Leakage	Reduction of perovskite material, sintering effects due to burning
Cathode contact layer	Incorporation of foreign elements	Change in materials properties (conductivity, CTE...)
Anode and substrate	Leakage	Reoxidation of Ni (microstructural changes leading to stresses and cracking)
Interconnect	Corrosion	Formation of less-conducting oxides, spallation of oxide layers, "breakaway corrosion"
Anode	Incorporation of foreign elements	Sulphur poisoning, carbon deposition, "metal dusting"
Raw materials	Foreign elements	e.g. Si, alkaline

Besides stack testing, the testing of stack components (e.g. cells sealed to a metallic frame), individual parts (single cell tests, sealant testing, corrosion testing of ICs) and the characterization of basic material parameters complete the various possible tests for understanding ageing, separating degradation influencing parameters and quantifying degradation phenomena.

In this paper, two degradation phenomena are presented in more detail: on the one hand, the incorporation of volatile chromium species into a cathode based on lanthanum-strontium-manganite (LSM) and the microstructural evolution of the layer protecting against chromium evaporation (PL) and the cathode contact layer (CCL) during operation. Both results are discussed on the basis of the results of stack testing (2-cell, so-called F-design stacks with 100x100mm² cells).

STACK COMPONENTS

The results presented here were obtained using stacks with the following characteristics and operating conditions:

- Cells: Ni/8YSZ substrate, Ni/8YSZ anode, 8YSZ electrolyte, LSM/8YSZ cathode, LSM current collector. 2 cells per stack
- Sealant: glass-ceramic based on B-Ca-Si-Al-glass
- Interconnect: Crofer22APU; FZJ F design, 2 planes
- Anodic contacting: Ni felt
- Cathodic contacting: CCL based on La-Mn-Cu-Co perovskite; PL based on MnO_2

- <u>Fuel</u>: 3% humidified hydrogen or methane + steam (S/C=2)
- <u>Oxidant</u>: air
- <u>Fuel utilization</u>: varying from 8 to 40%
- <u>Temperature</u>: 800°C
- <u>Current density</u>: 0.3 or 0.5A/cm² during long-term measurements
- <u>Operation times</u>: for Cr poisoning tests 3000h; for microstructural investigations between zero and 8200h

CHROMIUM POISONING OF LSM CATHODE

The results obtained from the long-term testing of four different stacks by varying, on the one hand, the current density applied (0.3 or 0.5 A/cm²) and, on the other hand, the type of fuel used (hydrogen + steam or methane + steam) were part of the European RealSOFC project. Figure 1 gives an overview of the voltage behaviour over time for all four stacks (8 planes).

From Figure 1 it is obvious that the fuel gas used has no influence on stack degradation, while the current density applied shows a distinct influence on degradation over the whole period of operation (~ 3000h). While the stacks operated at 0.3A/cm² show a quasi-linear degradation with an increase of the ASR of about 0.02 Ωcm² per 1000h, those operated at 0.5A/cm² reveal a progressive ASR increase over time.

After testing, all four stacks were carefully dissected plane by plane and various samples were taken from each plane for either microstructural investigations by SEM or for chemical analysis to quantify the amount of foreign elements incorporated in the cathode [1].

SEM characterization on polished cross-sectional samples shows that the cathode microstructure at the electrolyte/cathode interface has changed. It is three-phased (8YSZ, LSM and a foreign newly formed phase) and denser than that towards the current collection layer (Fig. 2). As an example, Figure 3 shows the microstructure of a stack operated at 0.5A/cm².

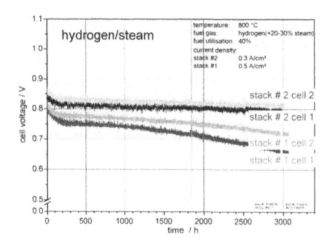

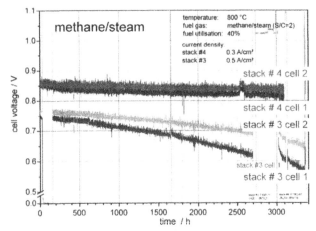

Fig. 1: Voltage-time plots of four stacks for Cr poisoning characterization; top: stacks operated under hydrogen, bottom: stacks operated under methane

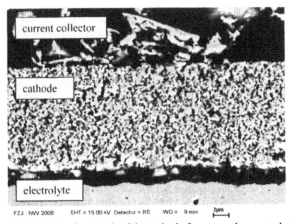

Fig. 2: SEM micrograph of the cathode from a stack operated at $0.3 A/cm^2$ under H_2 for 3000h

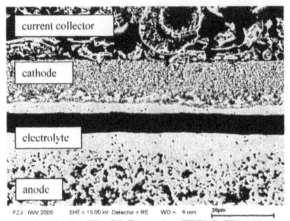

Fig. 3: SEM micrograph of the cathode from a stack operated at 0.5A/cm² under CH₄ for 3000h

By comparing the two micrographs, it is obvious that the microstructurally changed zone within the cathode is thicker in those stacks operated at higher current densities than in those operated at lower current densities.

EDX (energy-dispersive X-ray) point and area analysis found that the foreign phase formed consists of chromium (from the interconnect oxide layers), manganese (from the cathode), copper (from the CCL) and oxygen. It was concluded that a spinel phase is formed (calculations take into account the measured amounts of elements and the phase stabilities support these findings). These results are supported by literature data [4, 5].

Additionally, it was found that the changed zone is in most cases thicker at the air outlet than at the air inlet. This finding suggests that the air stream applied has a distinct influence on the transportation of the chromium species (as stated by Hilpert et al. and Yokokawa et al. [2, 3] mostly $CrO_2(OH)_2$ in humid air conditions). Therefore one significant diffusion path for the Cr from the interconnect oxide layers on the interconnect must be by gas phase diffusion. Whether the surface diffusion path also plays a significant role has been a matter of contention in the literature to date. Also the diffusion path of Cu is still unclear (gas phase or surface).

Besides the SEM characterization, the cathode was etched carefully (by a mixture of HCl and H_2O_2) and the dissolved material was characterized by ICP-OES (inductively coupled plasma with optical emission spectroscopy). This was to give additional information on whether the amount of foreign elements incorporated into the cathode is the reason for the differences in the microstructurally changed zone due to the fact that in porous structures only qualitative analysis is possible with SEM-EDX (not quantitative analysis).

The amounts of Cr and Cu analysed were calculated into amounts per square centimetre of cathode area. It was found that amounts between 0.2 and 0.8 µg/cm² can be analysed for either Cr or Cu. There was no clear trend towards higher amounts of Cr and Cu in those stacks operated at the higher current densities, but a tendency to higher amounts of the foreign elements at the air-out side of the stack was confirmed. Calculating the Cr:Cu ratio, it was found that at those stacks operated at 0.5A/cm² the amount of Cr was always higher than the amount of Cu, while for the stacks operated at the lower current densities there was a scattering of results. From the results obtained so far some conclusions can be drawn:

- There is no direct correlation between the amount of foreign elements within the cathode (especially Cr and Cu) and the degradation of the stack; there is only a change of the ratio Cr:Cu, and whether this influences the degradation rate is still unknown.

Concerning chromium incorporation, four main aspects may function as a degradation source in parallel (and their quantitative individual effect is still unclear): a) formation of a foreign (non-catalytic) phase (spinel), b) physical blocking of pores due to foreign phase formation, c) electrochemical blocking of the triple-phase boundaries and d) Mn depletion of the cathode (resulting in changed characteristics of the cathode material).

- The microstructurally changed zone within the cathode at the interface to the electrolyte is the thicker the higher the applied current density is and its thickness increases in parallel to the air stream.
- SEM micrographs and chemical analysis are in good agreement concerning the structural change of cathode and amount of Cr and Cu analysed.
- The foreign phase formed during Cr and Cu incorporation is a spinel.
- The influence of the foreign elements is complex: a) formation of a foreign phase with materials characteristics differing from those of the cathode (CTE, electrical conductivity etc, b) depletion of manganese from the LSM thus changing the characteristics of the cathode material, c) electrochemical blocking of the TPBs and d) physical blocking of the TPBs; additionally, other degradation-influencing factors may overlay the degradation caused by poisoning of the cathode by Cr and Cu.
- Up to now it is still unclear whether the Cu has a positive (less degradation), negative (higher degradation) or no influence on the degradation measured and caused only by chromium.

MICROSTRUCTURAL INVESTIGATION OF PL AND CCL

In using metallic interconnects containing high amounts of chromium and perovskite cathodes (LSM or LSFC) it is necessary to reduce, on the one hand, the polarization losses between cathode and interconnect and, on the other hand, to minimize chromium evaporation. Both objectives can be realized by introducing a chromium evaporation protection layer (PL) and a cathode contact layer (CCL). For the FZJ stacks, manganese oxide is used as the PL and a CCL based on a La-Mn-Cu-Co perovskite is applied to reduce polarization losses. Both layers were deposited by wet powder spraying (WPS) using alcoholic suspensions based on powders with particle sizes of less than 1 μm diameter. The layers are deposited on all surfaces in the stack in contact with air. There is no added sintering step, but the layers were dried and subsequently during start-up of the stack tempered at the same temperatures needed to crystallize the glass sealant. Additionally, the CCL ensures contacting during vitrification of the glass sealant powder and therefore acts as a kind of "elastic" contacting medium. Figure 4 shows the assembly of the components.

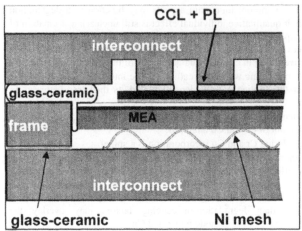

Fig. 4: Schematic cross-section of a repeating unit

As the interconnect design is based on rectangular channel structures, WPS was chosen as the coating technology. With WPS all areas, i.e. the top of the channels, the bottom and the perpendicular flanks, can be covered. It is usual for the envisaged thickness of the CCL (approx. 100-150µm) to differ between the three areas noted above. Figure 5 shows an example in the as-sprayed state. The thickness of the CCL at the flanks and the bottom of the channels is thinner than at the top and edges. Additionally, it is shown that the thickness at the top is not uniform but is concave curved. Especially in magnification D1, the two layers can be readily seen . Near the interface to the interconnect there is a thin manganese oxide PL above the thicker CCL. Besides the variations in thickness depending on the covered area, the CCL layer is porous (as desired for optimized gas transport), homogeneous and covers all necessary metallic areas.

The layer microstructure changes after a stack has been operated for 3020h at 800°C at a current density of 300mA/cm². Especially the manganese oxide layer is coarsened and therefore the real contact area is diminished (Fig. 6, 7; compare esp. pictures "D1" in Fig. 6 and complete Fig. 7). Additionally an in situ oxide layer formed on the steel can be seen. This oxide layer has a thickness of 2-4 µm and is composed of a chromia and a Cr-Mn-spinel layer with varying compositions [6, 7]. In Fig. 7, an SEM image of the same cell is presented. The coarsening of the PL and the unchanged microstructure of the CCL can be seen clearly. EDX analysis shows that chromium was incorporated into the CCL after operation (measured but not shown here), but this does not seem to influence the microstructure of the layer.

Prolonging stack operation time to more than 8000h leads in part to spallation of the layers after disassembly, indicating that the PL layer has become more brittle and porous. Spallation mostly takes place at the gas manifold structures. Continuous operation of the stack could presumably only be achieved due to the external mechanical load on the stack avoiding the detachment of the ceramic layers from the interconnect. The $Cr_2O_3/(Cr,Mn)_3O_4$ layers had nearly the same thickness as after 3000 h of operation.

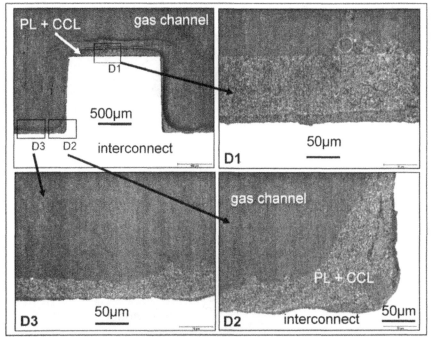

Fig. 5: Cross-section (optical microscopy) of CEPL and CCL after coating; top left: overview, top right: top of channel, bottom left: bottom of channel, bottom right: channel edge

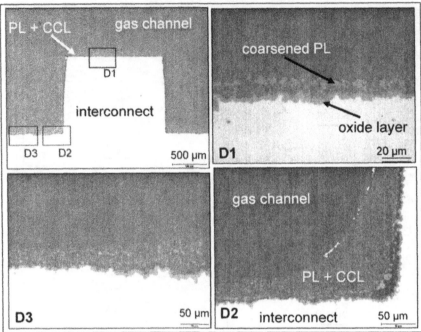

Fig. 6: Cross-section (optical microscopy) of PL and CCL taken from a stack after 3020 h of operation; top left: overview, top right: top of channel, bottom left: bottom of channel, bottom right: channel edge (please note the different magnifications due to varying layer thicknesses)

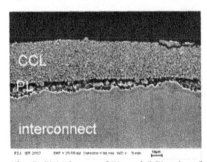

Fig. 7: SEM image of PL and CCL taken from a stack after operation for 3020 h. The PL shows strong grain coarsening and increased porosity whereas the CCL does not show any microstructural changes but rather the incorporation of Cr.

For long-term stack operation either the materials used as CCL or the coating technology must be changed. Novel materials are needed with comparable or even better physical parameters but with enhanced adhesion behaviour with respect to the interconnect. In particular, materials are required which react during stack start-up with the oxide layers formed on the interconnect or with the current collector of the cell. On the other hand, coating technologies must be developed which lead to dense layers (for the PL) or to layers with more uniform thickness in all areas which need to be covered. Possible coating techniques may be galvanic coating and subsequently oxidation, screen printing, roller coating or a dispenser system for the CCL on the contact area and only WPS for the protection layer. Other technologies such as physical vapour deposition, plasma spraying, and high velocity oxyfuel deposition are also possible.

The results obtained so far show that mid-term stack operations of up to more than 8000h of operation are possible, but in some cases degradation may occur with respect to the incorporation of foreign elements into the electrodes or the changing of microstructures and compositions of PL and CCL which may cause degradations with different time axes.

SUMMARY

SOFC stacks suffer from degradation during operation. The lowest degradation rates using metallic interconnects published up to now are in the range of 1% voltage loss per 1000 hours of operation. For auxiliary power unit applications these rates may be tolerable, but for stationary applications (> 40,000h of operation) these voltage losses (ASR increase) are not. Therefore numerous researchers all over the world are trying to understand the degradation mechanisms during stack or system operation.

Based on stack tests including metallic chromium-containing interconnects and cells with LSM cathodes, two examples of degradation causes are presented: on the one hand, the poisoning of the LSM cathode by volatile chromium vapour species and, on the other hand, the contact loss due to microstructural changes within the PL and the CCL. The coarsening of the PL and the incorporation of additional elements can influence the contacting within a stack (and therefore also the polarization losses) and also the intrinsic physical parameters of the layers and the Cr evaporation rates. Additionally, Cr incorporation into the LSM cathode leads to a reduction of triple-phase boundaries, manganese depletion of the LSM and physical filling of the pores and therefore to reduced cathode activity.

Further R&D to separate the influencing factors is needed in future so that single parameters can be better described thus additionally leading to a mathematical model of degradation phenomena in SOFCs.

ACKNOWLEDGEMENTS
The authors gratefully acknowledge SEM sample preparation by M. Kappertz, SEM support by Dr. D. Sebold and financial support contributed by the European Commission as part of the RealSOFC project.

REFERENCES

[1] Menzler N.H., de Haart L.G.J., Sebold D.: Characterization of cathode chromium incorporation during mid-term stack operation under various operational conditions. ECS Transactions 7 (1) (2007), 245-254
[2] Hilpert K., Das D., Miller M., Peck D.H., Weiß R.: Chromium vapor species over solid oxide fuel cell interconnect materials and their potential for degradation processes. J. Electrochem Soc. Vol. 143 No. 11 (1996), 3642-3647

[3]   Yokokawa H., Horita T., Sakai N., Yamaji K., Brito M.E., Xiong Y.-P., Kishimoto H.:
        Thermodynamic considerations on Cr poisoning in SOFC cathodes. Solid State Ionics 177
        (2006), 3193-3198
[4]   Jiang S.P., Zhang J.P., Zheng X.G.: A comparative investigation of chromium deposition
        at air electrodes of solid oxide fuel cells. J. Europ. Ceram. Soc. 22 (2002), 361-373
[5]   Paulson S.C., Birss V.I.: Chromium poisoning of LSM-YSZ SOFC cathodes I. J.
        Electrochem. Soc. 151 [11] (2004), A1961-A1968
[6]   Huczkowski P., Christiansen N., Shemet V., Niewolak L., Piron-Abellan J., Singheiser L.,
        Quadakkers W.J.: Growth mechanisms and electrical conductivity of oxide scales on
        ferritic steels proposed as interconnect materials for SOFCs. Fuel Cells 06 No. 2 (2006),
        93-99
[7]   Huczkowski P., Christiansen N., Shemet V, Piron-Abellan J., Singheiser L., Quadakkers
        W.J.: Oxidation induced lifetime limits of chromia forming ferritic interconnector steels.
        Transactions of the ASME Vol. 1 (2004), 30-34

PRESSURE AND GAS CONCENTRATION EFFECTS ON VOLTAGE *VS.* CURRENT CHARACTERISTICS OF A SOLID OXIDE FUEL CELL AND ELECTROLYZER

V. Hugo Schmidt and Laura M. Lediaev
Department of Physics, Montana State University
Bozeman, MT 59717

ABSTRACT
Effects of pressure and feed gas composition on voltage *vs.* current characteristics of a solid oxide fuel cell are calculated for operation in both the fuel cell and electrolyzer modes. Benefits of increasing pressure are considerable in going from 1 to 3 bars, but show diminishing returns as pressure is increased to 10 bars. Good agreement is found with experiment results obtained by another group at one atmosphere for H_2O/H_2 feed ratios of 15/85, 50/50, and 80/20.

INTRODUCTION
Pressurizing the input gases for SOFC's can improve performance. To quantify this effect, we look at relevant parts of our model for terminal voltage V as a function of electrolyte current density i.[1] Important inputs to this model are the fuel and exhaust gas concentrations at the anode-electrolyte interface. These depend on the anode pore tortuosity τ, which we determined[2] from a series of experiments by Jiang and Virkar[3] (JV). Our values for various fuel-exhaust and fuel-diluent-exhaust combinations used by JV centered on $\tau = 2.3$. This value is only slightly smaller than the middle of their τ range, taking into account that what they called τ is actually the "tortuosity factor" τ^2. In the present work, we use this $\tau = 2.3$ value and much of the development of those papers[2] in conjunction with our $V(i)$ model[1] for presenting two comparisons. One is a comparison of model predictions with JV results at 1 bar pressure for H_2O/H_2 feed gas molecular ratios of 15/85, 50/50, and 80/20. Surprisingly good agreement is found. The other comparison is for model predictions for these same three molecular ratios for three different operating pressures, 1 bar, 3 bars, and 10 bars. Even for this nearly exponential spacing of pressures, the benefits per pressure step go down as pressure increases, for operation both in the fuel cell and electrolyzer modes.

MODEL FOR VOLTAGE AS FUNCTION OF CURRENT DENSITY
Gas molecules from the anode and cathode pores are assumed to interact directly at the *triple phase boundary* (TPB) sites, located on lines where pore, electrode, and electrolyte meet. The SOFC with H_2 fuel and O_2 oxidizer has two reversible reactions, shown respectively for the anode and cathode below, together with net thermal energies liberated by the forward reactions. V_a and V_c are voltage steps at the electrolyte interfaces with the anode and cathode respectively, and do not include ohmic polarization.

$$H_2 + O^{2-} \leftrightarrow H_2O + 2e^-, \ U_a - qV_a; \tag{1}$$
$$\tfrac{1}{2}O_2 + 2e^- \leftrightarrow O^{2-}, \ U_c - qV_c. \tag{2}$$

Adding the two reactions and thermal energies yields

$$H_2 + \tfrac{1}{2}O_2 \leftrightarrow H_2O, \ U_a + U_c - q(V_a + V_c) \equiv U - q(V + V_{ohm}), \tag{3}$$
$$U = U_a + U_c. \tag{4}$$

Here, U is heat of reaction for combustion of hydrogen (or carbon monoxide), q is twice the proton charge, V is terminal voltage and V_{ohm} is ohmic polarization.

We summarize now the model[1] for the anode-electrolyte reaction of Eq. (1). Consider oxygen ion sites on the electrolyte surface along a TPB line. Let v_a be the *vacancy probability* that a given O^{2-} site is vacant, so $1-v_a$ is the probability that it is occupied. We designate as v_f (**f** subscript for incoming fuel molecule) the *anode forward reaction rate*, the probability per unit time that the forward reaction in Eq. (3) will occur at each occupied site. Similarly we designate as v_e (**e** subscript for incoming exhaust molecule) the *anode reverse reaction rate* that we assume can occur only at unoccupied sites.

The reaction rate v_f is the *attempt rate* v_{fo} multiplied by the *success probability* that the forward reaction of Eq. (1) will occur when the molecule strikes. To find the attempt rate, consider a fuel gas concentration n_{fi} at the interface. Half of the fuel molecules will be traveling toward the surface with mean normal velocity component $<u_y>$, so the flux of molecules striking the surface will be $\frac{1}{2}n_f<u_y>$. Since $\frac{1}{2}kT=\frac{1}{2}m_f<u_y^2>$, $<u_y> \cong (kT/m_f)^{1/2}$, where m_f is the fuel molecule mass, so the flux is $\frac{1}{2}n_f(kT/m_f)^{1/2}$. For a reaction site of area s, the impingement rate is

$$v_{fo}=\tfrac{1}{2}n_f s(kT/m_f)^{1/2}. \tag{5}$$

A similar expression, but with subscript $_f$ replaced by $_e$, holds for the H_2O exhaust molecule impingement rate.

The success probability depends partly on the heat of formation U_a released if the reaction occurs. However, the two electrons of combined charge magnitude q must jump "uphill" across a potential step of magnitude V_a to reach the Ni electrode. Thus the net energy required for the reaction to occur is qV_a-U_a. If it does not occur, the fuel molecule bounces back into the anode pore.

This reaction attempt is a *collision event* whose probability of either final state (reaction, or no reaction) is assumed to be governed by classical (Boltzmann) statistics. The un-normalized probabilities for the two final states to occur then are 1 for no reaction and $\exp[-(qV_a-U_a)/kT]$ for a reaction. Note that we assume only an energy step, but not a barrier between two minima corresponding to initial and final states. Normalizing, and multiplying by v_{fo}, we obtain reaction rate per O^{2-} site

$$v_f=v_{fo}\exp[-(qV_a-U_a)/2kT]/2\cosh[(qV_a-U_a)/2kT]. \tag{6}$$

A similar expression, but with opposite sign for the argument of the exponential, holds for the reverse reaction based on incoming exhaust molecules:

$$v_e=v_{eo}\exp[(qV_a-U_a)/2kT]/2\cosh[(qV_a-U_a)/2kT]. \tag{7}$$

For open circuit, forward and reverse reaction rates are equal because $i=0$ in the electrolyte. The equivalent equal and opposite forward and reverse current densities in the electrolyte are designated in the literature as i_0, the *exchange current density*. In general, i_0 will have different values, i_{0a} and i_{0c}, for the anode and cathode reactions.

We make the approximation that the oxygen reaction site areas cover the whole surface but do not overlap, so that the number of sites per cm^2 is $1/s$. Only a fraction f_a of oxygen sites, to be estimated later, will be at anode interface TPB's and thus available for reaction. The current densities are related by a factor q to the gas flow densities resulting in reactions. Putting all the factors together, we obtain

$$i_f = \tfrac{1}{2} n_{fi}(kT/m_f)^{1/2} q(1-v_a) f_a e^{\alpha} /2\cosh\alpha \equiv a e^{\alpha} /2\cosh\alpha, \tag{8}$$

$$i_c = \tfrac{1}{2} n_{ci}(kT/m_c)^{1/2} q v_a f_a e^{-\alpha} /2\cosh\alpha \equiv b e^{-\alpha} /2\cosh\alpha, \tag{9}$$

$$\alpha \equiv (U_a - qV_a)/2kT, \tag{10}$$

$$a = \tfrac{1}{2} n_{fi}(kT/m_f)^{1/2} q(1-v_a) f_a, \tag{11}$$

$$b = \tfrac{1}{2} n_{ci}(kT/m_e)^{1/2} q v_a f_a. \tag{12}$$

Here, a and b respectively are the forward and reverse anode *attempt current densities*. We designate open-circuit values below with subscript $_0$. Then the anode exchange current density i_{0a} is

$$i_{0a} = i_{f0} = i_{c0}. \tag{13}$$

Rearranging Eqs. (8) and (9) and using

$$i = i_f - i_c, \tag{14}$$

we obtain

$$i = \tfrac{1}{2}(a-b) + \tfrac{1}{2}(a+b)\tanh\alpha, \tag{15}$$

Note that Eq. (15) requires $-b < i < a$, which means i cannot exceed these limits in the electrolysis and fuel cell modes respectively.

A similar range restriction applies to the cathode-electrolyte interface. These ranges are

$$-b \le i \le a \text{ (anode limitation)}, \quad -d \le i \le c \text{ (cathode limitation)}. \tag{16}$$

We now find expressions for c and d. The cathode forward attempt current density c is given by

$$c = \tfrac{1}{2} n_{ci}(kT/m_c)^{1/2}(2q) v_c^2 f_c. \tag{17}$$

It differs from the expression for a in that twice the charge $(2q)$ is transferred. The v_c^2 factor is the probability that both oxygen atoms of the impinging O_2 molecule strike vacant sites; otherwise it is assumed that the molecule bounces back into the pore. The f_c factor denotes the probability that the O_2 molecule strikes a TPB.

The calculation of d is different than for a, b, and c because the reaction rate is not limited by an incoming gas flow rate. Instead, d depends on a typical solid-state reaction attempt frequency $v = 10^{13}$/s and on the density of adjacent occupied oxygen ion site pairs located on TPB lines. To find this density we assume a (001) surface orientation with oxygen site density $1/s = 1.44 \times 10^{15}$/cm^2. A given oxygen site has 8 nearest surface neighbors along <100> and <110> directions, but only a fraction $(1-v_c)$ (assumed to be 0.3 as discussed below) of these will be occupied. Multiplying the above factors together, the density of adjacent oxygen ion pairs is $\tfrac{1}{2} s^{-1} 8(1-v_c)^2$, where the $\tfrac{1}{2}$ factor eliminates double counting. Then d is the foregoing factor multiplied by $2qv f_c$, where $q = 3.2 \times 10^{-19}$ C and $f_c = 4.54 \times 10^{-4}$ as will be found below, so

$$d=d_0=8qv\,s^{-1}(1-v_c)^2f_c=1.51\times10^6 \text{ A/cm}^2. \tag{18}$$

We now use c and d in the following five equations for the cathode interface, which are the equivalents of Eqs. (8-10,14,15) for the anode interface.

$$i_h=ce^{\gamma}/2\cosh\gamma, \tag{19}$$
$$i_g=de^{-\gamma}/2\cosh\gamma, \tag{20}$$
$$\gamma\equiv2(U_c-qV_c)/2kT, \tag{21}$$
(the first 2 is due to double-size charge transfer in the cathode reaction)
$$i=i_h-i_g\ (=0 \text{ for open circuit}), \tag{22}$$
$$i=\tfrac{1}{2}(c-d)+\tfrac{1}{2}(c+d)\tanh\gamma, \tag{23}$$

From Eq. (15) we have

$$\tanh\alpha=(2i-a+b)/(a+b), \tag{24}$$
$$\alpha=\tfrac{1}{2}\ln[(b+i)/(a-i)]=(U_a-qV_a)/2kT, \tag{25}$$

where the last expression in Eq. (25) comes from Eq. (10) . Solving Eq. (25) for V_a gives

$$V_a=U_a/q-(kT/q)\ln[(b+i)/(a-i)]. \tag{26}$$

Similarly, V_c is found from Eqs. (21) and (23) to be

$$V_c=U_c/q-(kT/q)\ln[(d+i)/(c-i)]^{1/2}. \tag{27}$$

To find the terminal voltage

$$V=V_0-V_{act}-V_{conc}-V_{ohm}=V_a+V_c-V_{ohm}. \tag{28}$$

we use Eqs. (4) and (26-28) to obtain

$$V=U/q-(kT/q)\{\ln[(b+i)/(a-i)]+\ln[(d+i)/(c-i)]^{1/2}\}-V_{ohm}$$
$$=U/q-(kT/q)\ln[(b+i)(d+i)^{1/2}/(a-i)(c-i)^{1/2}]-V_{ohm}. \tag{29}$$

The open-circuit emf V_0, from Eq. (29) with $i=V_{ohm}=0$, and with a, b, c, d set at their open-circuit values, is

$$V_0=U/q-(kT/q)\ln(b_0d_0^{1/2}/a_0c_0^{1/2}). \tag{30}$$

GAS CONCENTRATIONS AT INTERFACES VS. CURRENT DENSITY

Careful treatment of a, b, and c as functions of i and other parameters in Eq. (29) requires a detailed analysis of fuel, exhaust, and oxidant gas concentrations n_{fi}, n_{ei}, and n_{ci} at the electrode-electrolyte interfaces, so a separate section is devoted to this analysis.

This analysis begins with the dusty-gas model which includes effects of Knudsen diffusion (molecular collisions with electrode pore walls) as well as binary collisions between unlike molecules. It does not include laminar or turbulent flow effects which are unimportant for the small diameter pores typically used in SOFC electrodes. The basic equation for this model when converted to molecular units[2] is

$$J_i/D_{Ki}+(J_i n_j - J_j n_i)kT/D_{ij1}P_1=-\partial n_i/\partial x. \tag{31}$$

Here, J_i is the molecular flux rate (molec/cm^2-s) of the ith gas species, D_{ij1} is the binary diffusion coefficient at $P_1 \equiv 1$ atmosphere, and the nkT-type terms in the numerator divided by P_1 convert D_{ij1} to D_{ij} at the actual total pressure at any position x along the anode pore. The variable x is distance along the pore, so the total length L of the pore is $L=\tau w$, where τ is the anode pore tortuosity and w is the thickness of the electrode. The Knudsen diffusion coefficient D_{Ki} is given by

$$D_{K,i} = \tfrac{2}{3}(8kT/\pi m_i)^{1/2}\bar{r} , \tag{32}$$

where $\bar{r}$ is the mean pore radius.

For the binary fuel/exhaust gas systems H_2/H_2O and CO/CO_2, if subscript $_f$ is for the fuel gas and $_e$ is for the exhaust gas, $J_e=-J_f$ in Eq. (3). Here, $J_f=i\tau/\phi q$, where i is the current density $(A/cm^2=C/cm^2 s)$ in the solid electrolyte and q is the charge $(3.2\times10^{-19}$ C/molecule) carried per gas molecule annihilated/created in the reaction at the anode/electrolyte interface.

The τ/ϕ factor is an enhancement factor by which the flow density J_i is enhanced compared to its value if the anode were completely porous ($\phi=\tau=1$). To derive this result, note that this enhancement factor for an anode of area S with N_{pa} pores each of cross-sectional area A_x is $S/N_{pa}A_x$. The porosity ϕ is the total pore volume $N_{pa}A_x\tau w$ divided by the anode volume Sw, so $\phi=N_{pa}A_x\tau/S$. By rearranging this equation we find that the enhancement factor $S/N_{pa}A_x$ is τ/ϕ.

Eq. (31) for the fuel gas can now be written, after setting $n_e+n_f=n$, as

$$\partial n_f/\partial x= -(i\tau/\phi q)(1/D_{Kf}+nkT/D_{ef1}P_1)\equiv -c_3\tau -c_4 n\tau , \tag{33}$$

where c_3 and c_4 are positive constants for fuel cell operation, and negative constants (because i is then negative) for electrolyzer operation. The subscript $_f$ refers to the fuel gas (H_2 or CO) in the binary gas system. The corresponding equation for the exhaust gas, after defining $(m_e/m_f)^{1/2}\equiv 2h+1$, is

$$\partial n_e/\partial x=(2h+1)c_3\tau +c_4 n\tau . \tag{34}$$

Adding Eqs. (33) and (34) yields the equation for n and its solution,

$$\partial n/\partial x=2hc_3\tau , \quad n=p_a n_1+2hc_3\tau x. \tag{35}$$

where p_a is dimensionless pressure in atmospheres and n_1 is total gas molecular concentration at 1 atmosphere. The ideal gas law, $P=nkT$, shows that for fuel cell operation the total pressure P increases linearly with x in traversing the anode from the plenum to the solid electrolyte. For electrolyzer operation, P decreases linearly with x.

Now that $n(x)$ is known from Eq. (35), we can solve Eqs. (33) and (34) for n_f and n_e.

$$\partial n_f/\partial x=-c_3\tau -c_4(p_a n_1+2hc_3\tau x)\tau , \tag{36}$$
$$n_f=n_{fp}-(c_3+c_4 p_a n_1)\tau x-hc_3c_4\tau^2 x^2,$$

$$\partial n_e / \partial x = (2h+1)c_3 \tau + c_4(p_a n_1 + 2hc_3 \tau x) \tau,$$ (37)
$$n_e = n_{ep} + [(2h+1)c_3 + c_4 p_a n_1] \tau x + hc_3 c_4 \tau^2 x^2.$$

Here n_{fp} and n_{ep} are respectively the concentrations of fuel and exhaust gas in the plenum. Even though Eq. (37) for n_c has a positive term quadratic in x, Eq. (34) insures that its slope remains negative in the electrolyzer mode, because n is positive within the anode which is the region of physical interest.

The next step is finding expressions for n_{fp} and n_{ep}. For open circuit they are $p_f p_a n_1$ and $p_e p_a n_1$ respectively for the fuel and exhaust gases. Here p_f and p_e respectively are the fractions of fuel and exhaust molecules in the anode feed gas, so $p_f + p_e = 1$ and the fuel/exhaust molecular ratio flowing out the anode plenum exit port is p_f / p_e. For closed circuit there will be fuel and exhaust gases flowing between the plenum and the anode pores in addition to the feed gas flowing into the plenum, so the fuel/exhaust ratios in the plenum and exit port will change. We assume good mixing so that these ratios are the same. This ratio must be (net fuel flow into plenum)/(net exhaust flow into plenum) considering only the feed gas and flow in or out of the anode, not out of the exit port. We are only interested in a ratio, so can convert molecular flow rates into equivalent currents. Thus we can let $-i$ and i represent the fuel flow into the anode, and exhaust flow out of the anode, respectively. (This magnitude equality holds for H_2/H_2O and CO/CO_2 fuel/exhaust combinations, but not for some other combinations.) Given a known metered total molecular anode feed flow rate $p_f j_1 + p_e j_1 = j_1$ molec/s and using the conversion $i_1 = j_1 q / S$, where S is electrolyte cross-sectional area, we have net fuel/exhaust inflow ratio

$$(p_f i_1 - i)/(p_e i_1 + i) = n_{fp}/n_{ep}.$$ (38)

For operation at p_a atmospheres, we require $n_{fp} + n_{ep} = p_a n_1$, so we obtain

$$n_{fp} = (p_f - i/i_1)p_a n_1, \quad n_{ep} = (p_e + i/i_1)p_a n_1.$$ (39)

Making these substitutions in Eq. (36), we obtain

$$n_{fi} = (p_f - i/i_1)p_a n_1 - (iw \tau^2 / \phi q)(1/D_{Kf} + kTn_1 p_a / P_1 D_{efl}) - i^2 hkTw^2 \tau^4 / \phi^2 q^2 P_1 D_{Kf} D_{efl}$$
$$\equiv p_f p_a n_1 - C_1 i - C_2 p_a i - C_3 i^2.$$ (40)
$$C_1 = (w \tau^2 / \phi q)(1/D_{Kf}).$$
$$C_2 = n_1/i_1 + (w \tau^2 / \phi q)(kTn_1/P_1 D_{efl}).$$
$$C_3 = hkTw^2 \tau^4 / \phi^2 q^2 P_1 D_{Kf} D_{efl}.$$

We also need

$$n_{ei} = n_i - n_{fi} = p_a n_1 + 2ihw \tau^2 / \phi q D_{Kf} - n_{fi} \equiv p_e p_a n_1 + C_4 i + C_2 p_a i + C_3 i^2.$$ (41)
$$C_4 = (2h+1)(w \tau^2 / \phi q)(1/D_{Kf}).$$

MODEL PARAMETER NUMERICAL VALUES FOR JIANG-VIRKAR APPARATUS

We will make model predictions using temperature and feed gas parameters and cell dimensions and characteristics employed by Jiang and Virkar[3]. We will compare predictions with their experimental results where available, and will also extend predictions into pressure ranges higher than atmospheric and into the electrolyzer mode of operation.

In their cell, the anode and anode interlayer were Ni + YSZ (yttria-stabilized zirconia), the electrolyte was a YSZ + SDC (samaria-doped ceria) bilayer, the cathode interlayer was LSC (Sr-doped LaCoO₃) + SDC, and the cathode was LSC.

For the anode attempt current density a, we find numerical values for parameters in Eq. (11), repeated below.

$$a=\tfrac{1}{2}n_{fi}(kT/m_f)^{1/2}q(1-v_a)f_a, \tag{42}$$

JV performed all runs at atmospheric pressure $P_1=1.015\times10^5$ N/m² and at $T=1073$ K. To find n_{fi} and similar gas concentrations, we will need the total gas concentration n_1 at one atmosphere. From the ideal gas law, using $k=1.38\times10^{-23}$ J/K, we have

$$n_1=P_1/kT=6.855\times10^{18}\,\text{molec/cm}^3. \tag{43}$$

The mass m_f of the H_2 molecule is about 2 AMU or 3.32×10^{-27} kg. The charge q per reaction of Eq. (1) is 2 proton charges, or 3.20×10^{-19} C.

For both the anode-electrolyte and cathode-electrolyte interfaces, we do not assume that the vacancy concentrations v at the oxygen sites next to the TPB are the same as in the electrolyte bulk. Instead, we initially assumed $v=1-v=0.5$ for both interfaces. We later found that to get sufficient oxygen uptake, $v_c=0.7$ was needed at the cathode interface, but in previous work[1] we kept the value $v_a=0.5$ for the anode interface value. Since v_a and v_c are probably impossible to measure, they can be considered as adjustable constants. In reality, v_a and v_c may depend on i, but in this model they are assumed constant. To minimize the number of adjustable constants, in this work we set $v_a=v_c=0.7$.

To find both f_a and d, we need to find s and the site spacing $s^{1/2}$. To do this, we consider a (001) surface on the yttria-stabilized zirconia (YSZ) electrolyte with oxygen sites as the outer layer. These sites form a square array, with site spacing $s^{1/2}=2.64\times10^{-8}$ cm and area per site $s=6.97\times10^{-16}$ cm².

To find the fraction f_a of sites at anode interface TPBs, we must find d_t, the TPB length per cm² of electrolyte surface. We model the pores as circular with average radius $\bar{r}=0.5$ micron as found by JV[3]. We choose average pore tortuosity $\tau=2.3$, as found from our analysis of the JV data.[2] This tortuosity can be modeled as resulting from each pore being straight, but inclined at an angle $\cos^{-1}(1/\tau)$. The pore then meets the electrolyte at this angle and its intersection with the electrolyte is an ellipse of semimajor axis $\tau\bar{r}$ and semiminor axis $\bar{r}$. The pore area as it impinges on the electrolyte then is $\pi\tau\bar{r}^2$. The total pore area impinging per cm² of electrolyte is the porosity ϕ, found by JV to be $\phi=0.54$, so the number n_p of pores impinging per cm² of electrolyte is $n_p=\phi/\pi\tau\bar{r}^2=2.989\times10^7/\text{cm}^2$. Each pore perimeter P_p as it impinges on the electrolyte is, according to Marks' Handbook[4],

$$P_p=\pi\bar{r}(1+\tau)[1+\tfrac{1}{4}m^2+\tfrac{1}{64}m^4+\tfrac{1}{256}m^6+\cdots], \tag{44}$$

where $m=(\tau-1)/(\tau+1)$, so $P_p=5.753$ microns$=5.753\times10^{-4}$ cm. Multiplying by n_p yields $d_t=17196$ cm⁻¹. If we assume the TPB has width $s^{1/2}=2.64$ A°, the fraction f_a of anode-side electrolyte surface consisting of TPB is only $f_a=d_ts^{1/2}=4.54\times10^{-4}$. We have insufficient information for calculating the cathode-side f_c, so we assume $f_c=f_a$.

Multiplying all the factors in Eq. (42) except n_{fi} yields for H_2/H_2O feed gas the value

$a/n_{fi}{=}4.602x10^{-18}$ A-cm. $\qquad$ (45)

In finding b/n_{ei} for H_2/H_2O from Eq. (12), we note that in comparing with Eq. (11) the molecular mass is 9 times larger and $v_a{=}0.7$ replaces $(1{-}v_a){=}0.3$, so b/n_{ei} is $\frac{7}{9}a/n_{fi}$, or

$b/n_{ei}{=}3.580x10^{-18}$ A-cm. $\qquad$ (46)

Eq. (17) for the cathode forward attempt current density c, repeated here, is

$$c{=}\tfrac{1}{2}n_{ci}(kT/m_c)^{1/2}(2q)v_c^2f_c. \qquad (47)$$

To evaluate c/n_{ci}, we compare pertinent factors in Eq. (47) with related factors in Eq. (42) for a. In the square-root velocity factor, the mass ratio $m_f/m_c{=}2/32$ and its square root is $\frac{1}{4}$. The reaction $O_2{+}4e^-\leftrightarrow 2O^{2-}$ involves transfer of charge $2q$. The v_c^2 factor replacing the $(1{-}v_a)$ factor is the probability that both oxygen atoms of the impinging O_2 molecule strike vacant sites; otherwise it is assumed that the molecule bounces back into the pore. As discussed above, we chose $v_c{=}v_a{=}0.7$ and $f_c{=}f_a{=}4.54x10^{-4}$. The overall ratio is $\frac{1}{4}x2(0.49/0.3){=}49/60$, so

$c/n_{ci}{=}3.758x10^{-18}$ A-cm. $\qquad$ (48)

No pressure loss is assumed across the cathode, and a high enough air flow is assumed so that the 21 molecular % O_2 is not appreciably depleted even for large i. Then we have $n_{ci}{=}0.21p_an_1$, so

$c{=}5.410p_a$ A/cm^2. $\qquad$ (49)

In Eq. (18), repeated here, we found that

$d{=}d_0{=}8qvs^{-1}(1{-}v_c)^2f_c{=}1.51x10^6$ A/cm^2 $\qquad$ (50)

and is assumed independent of pressure, temperature, and current density.

Now we list the parameters necessary for calculating C_1, C_2, C_3, and C_4 in equations above for H_2/H_2O anode feed gas mixtures used by Jiang and Virkar.[3]

$n_1{=}6.855x10^{18}$/cm^3, $j_1{=}6.286x10^{19}$ molec/s, $S{=}1.1$ cm^2, $i_1{=}18.29$ A/cm^2, $\quad$ (51)
$w{=}0.11$ cm, $\tau{=}2.3$, $\phi{=}0.54$, $q{=}3.2x10^{-19}$ C, $D_{Kf}{=}11.3$ cm^2/s, $k{=}1.38x10^{-23}$ J/K, $T{=}1073$ K,
$P_1{=}1.015x10^5$ N/m^2, $D_{ef1}{=}7.704$ cm^2/s, $h{=}1$.

Using the above parameters, we obtain

$C_1{=}2.980x10^{17}$ s/cm-C, $C_2{=}8.119x10^{17}$ s/cm-C, $\qquad$ (52)
$C_3{=}1.900x10^{16}$ cm-s^2/C^2, $C_4{=}3C_1{=}8.94x10^{17}$ S/cm-C.

These parameters are needed to calculate n_{fi} and n_{ei} in Eqs. (40) and (41). Then, n_{fi} and n_{ei} can be inserted into Eqs. (45) and (46) to find a and b).

To find the U/q term in Eq. (29) for V, we need to find the effect of temperature on the gas-phase standard emf. This emf is listed as 1.18 V at 25 °C. The enthalpy changes in going from 25

$^{\circ}$C to 800 $^{\circ}$C, translated into volts emf, are -0.1498 V for H_2O, 0.1184 V for H_2, and 0.0656 V for $\frac{1}{2}O_2$, giving a standard emf U/q at 800 $^{\circ}$C of

$$U/q = 1.214 \text{ V}. \tag{53}$$

We also need the prefactor of the logarithmic term in Eq. (29), which is

$$kT/q = 0.0463 \text{ V}. \tag{54}$$

Finally, we subtract the ohmic polarization V_{ohm} to find the predicted terminal voltage V from Eq. (29). Comparing JV Fig. 12 for the total polarization without the ohmic contribution with their Fig. 5 that shows V for the H_2/H_2O runs,[3] we obtain

$$V_{ohm} = (0.122 \text{ ohm-cm}^2)i \tag{55}$$

PRESENTATION AND DISCUSSION OF RESULTS

The comparison of calculated values of voltage and of power density at 1 bar with experimental results of Jiang and Virkar[3] appears in Fig. 1. The agreement is quite satisfactory, especially considering that the model has only one adjustable parameter, namely the vacancy concentration of 0.7. A smaller value would extend the calculated curves to the right and improve the shown fits slightly, but would also cause cutoff from oxygen starvation for 100% H_2 feed gas (curve and data not shown) at current densities where such cutoff is not observed in the JV data. For operation as an electrolyzer, the straight line for the power density shows hydrogen production is proportional to power density in that range. This proportionality factor decreases where the voltage curves bend upward at larger negative current densities.

The remaining three Figures show the effects of increasing pressure for H_2O/H_2 ratios of 15/85, 50/50, and 80/20 respectively. The curves shift to the left with increasing steam concentration, as expected. For the 15/85 run, the power density increases from about 1.45 to about 1.7 W/cm^2 as pressure is increased from 1 to 3 bar, but it increases by a smaller amount to about 1.9 W/cm^2 at 10 bar. For lower H_2 content, the numerical increases in W/cm^2 with these increasing steps are smaller, but the fractional increases are larger.

Results not shown here indicate that benefits of higher pressure are greater if the ohmic and concentration polarization can be reduced, so long as the higher power outputs do not cause excessive heating.

Overall, we have reinforced the general opinion that operation at elevated pressure is desirable if added cost and complexity are not too great. Also, we have shown that our model gives good agreement with experimental results for operation at 1 bar for a particular SOFC apparatus.

ACKNOWLEDGMENTS

Helpful discussions with Larry Pederson and Stephen Sofie are gratefully acknowledged. This work was supported by DOE under subcontract DE-AC06-76RL01839 from Battelle Memorial Institute and PNNL.

REFERENCES

1. V. H. Schmidt, in *ECS Transactions*, vol. 6, issue 21, Design of Electrode Structures, E. De Castro, L. Lipp, and D. Mah, Eds., pp. 11-24 (2008).
2. V. H. Schmidt, C.-L. Tsai, and L. M. Lediaev, in N. P. Bansal (Ed.), *Advances in Solid Oxide Fuel Cells III*, Wiley, Hoboken, NJ, pp. 129-140 (2008).

3. Y. Jiang and A.V. Virkar, *J. Electrochem. Soc.*, **150**, A942-A951 (2003).
4. L. S. Marks, *Mechanical Engineers' Handbook*, p. 107, McGraw-Hill, New York (1941).

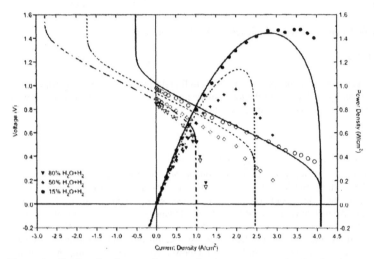

Figure 1. Predicted voltage and power density *vs.* current density for three ratios of steam to hydrogen feed gas, compared with experimental results of Jiang and Virkar.[3]

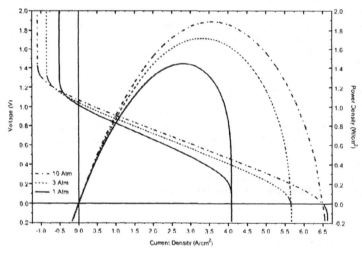

Figure 2. Predicted voltage and power density *vs.* current density for three operating pressures, for 15%/85% H_2O/H_2 feed gas molecular ratio.

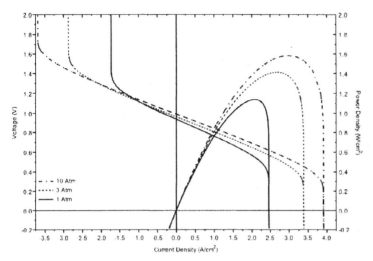

Figure 3. Predicted voltage and power density *vs.* current density for three operating pressures, for 50%/50% H_2O/H_2 feed gas molecular ratio

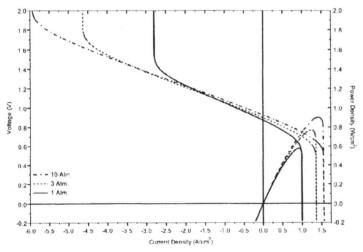

Figure 4. Predicted voltage and power density *vs.* current density for three operating pressures, for 80%/20% H_2O/H_2 feed gas molecular ratio.

IN-SITU TEMPERATURE-DEPENDENT X-RAY DIFFRACTION STUDY OF
Ba(Zr$_{0.8-x}$Ce$_x$Y$_{0.2}$)O$_{3-\delta}$ CERAMICS

C.-S. Tu[a], R. R. Chien[b], S.-C. Lee[a], C.-L. Tsai[b], V. H. Schmidt[b], A. Keith[c], S. A. Hall[d], and N. P. Santorsola[b]

[a]Department of Physics, Fu Jen Catholic University, Taipei 242, Taiwan
[b]Department of Physics, Montana State University, Bozeman, MT 59717, USA
[c]Department of Physics, University of North Dakota, Grand Forks, ND 58202, USA
[d]Baker Prairie Middle School, Canby, OR 97013, USA

ABSTRACT
    This work is a study of chemical stability of proton-conducting ceramics Ba(Zr$_{0.8-x}$Ce$_x$Y$_{0.2}$)O$_{3-\delta}$ (x=0.2, 0.3, and 0.4) in 1 atm pure CO$_2$–flowing atmosphere by using *in-situ* x-ray diffraction in the region of 25–1000 °C upon heating. Decomposition by the reaction, Ba(Zr$_{0.8-x}$Ce$_x$Y$_{0.2}$)O$_3$+CO$_2$→ BaCO$_3$+Ce(Zr,Y)O$_2$, was observed for x=0.3 and 0.4 at temperature above 550 °C. For x=0.20, decomposition was not observed up to the maximum measuring temperature 1000 °C. This study suggests that Ba(Zr$_{0.8-x}$Ce$_x$Y$_{0.2}$)O$_{3-\delta}$ ceramics for x≤0.2 are promising candidates for applications of proton-conducting solid-oxide fuel cells even in the intermediate temperature region of 600-900 °C.

INTRODUCTION
    One important issue facing protonic conducting ceramics for hydrogen purification and solid oxide fuel cell (SOFC) electrolytes is the chemical instability to the environment, especially reaction with coal syngas components such as CO$_2$, H$_2$S, and other trace species.[1,2] However, the effect of coal syngas species on SOFC related materials is not presently well known. Though proton conductors are promising candidates for SOFCs because of their low activation energy,[3-6] the major challenge for these materials is to find a proper compromise between ionic conductivity and chemical stability in various environments. For example, doped BaCeO$_3$ has been known to exhibit high ionic conductivity at high temperatures (≥500 °C) but poor chemical stability was observed in CO$_2$ and H$_2$O atmospheres.[3-9] On the other hand, yttrium-doped BaZrO$_3$ shows both sufficient chemical and thermal stability.[10,11] Zr-doping can improve chemical stability but decreases the ionic conductivity. Partially substituting Zr for Ce reduces its tendency to decompose into BaCO$_3$ and other oxides in CO$_2$–containing environment at intermediate temperatures (typically 700–850 °C) or high operation temperatures. Therefore, it has been a goal to find yttrium-doped Ba(Ce,Zr)O$_3$ proton conducting ceramics having both sufficient ionic conductivity and chemical stability by replacing a desired fraction of cerium in BaCeO$_3$ with Zr or using a processing method.[13-15]
    Recent post x-ray diffraction (XRD) results of Ba(Ce$_{0.8}$Y$_{0.2}$)O$_3$ (BCY20) and Ba(Zr$_{0.1}$Ce$_{0.7}$Y$_{0.2}$)O$_{3-\delta}$ (BZCY7) powders, suggested that BCY20 decomposed to BaCO$_3$, CeO$_2$ and Y$_2$O$_3$ after exposure to 2% CO$_2$ (with H$_2$) at 500 °C for one week.[12] However, BZCY7 remained unchanged in 2% CO$_2$–containing atmosphere at 500 °C and exhibited a good chemical and kinetic stability below 500 °C.[12] In a similar XRD investigation of the BZCY7 powders, the structure still remained the same after exposure to H$_2$ containing 15% H$_2$O for one week, implying that BZCY7 is kinetic stable at 500 °C in an atmosphere containing 15% water vapor.[12]
    In this work, *in-situ* temperature-dependent XRD was employed from room temperature to 1000 °C to investigate chemical decomposition in a 1 atm pure CO$_2$ flowing environment for Ba(Zr$_{0.8-x}$Ce$_x$Y$_{0.2}$)O$_{3-\delta}$ (x=0.2, 0.3, and 0.4) calcined powders. It was found that BaCO$_3$ and Ce(Zr,Y)O$_2$ were not detected below 550 °C for all components. However, BaCO$_3$ and Ce(Zr,Y)O$_2$ begin to develop above 550 °C for higher cerium components BZCY (x=0.3 and 0.4). BZCY (x=0.2)

powder samples exhibit an obvious chemical stability up to 900 °C.

EXPERIMENTAL

Yttrium-doped barium-zirconium-cerium-yttrium oxide powders Ba(Zr$_{0.8-x}$Ce$_x$Y$_{0.2}$)O$_{3-\delta}$ (x=0.2, 0.3, and 0.4) were synthesized by the Glycine-Nitrate process. BZCY synthesized powders used in this study were calcined at 1300 °C in air for 5 hours to ensure a cubic perovskite phase. In our preliminary x-ray results, BaCO$_3$ and Ce(Zr,Y)O$_2$ components were detected in powders calcined below 1100 °C for 5 hours. Hereafter, BZCY442, BZCY532, and BZCY622 represent Ba(Zr$_{0.4}$Ce$_{0.4}$Y$_{0.2}$)O$_{3-\delta}$, Ba(Zr$_{0.5}$Ce$_{0.3}$Y$_{0.2}$)O$_{3-\delta}$, and Ba(Zr$_{0.6}$Ce$_{0.2}$Y$_{0.2}$)O$_{3-\delta}$, respectively. For *in-situ* x-ray diffraction measurements, a high-temperature Rigaku Model MultiFlex x-ray diffractometer with Cu Kα_1 (λ=0.15406 nm) and Cu Kα_2 (λ=0.15444 nm) radiations was used. The intensity ratio between Kα_1 and Kα_2 is about 2:1.[16] Powders were placed and smoothly pressed on the platinum sample holder. The powder samples were scanned at room temperature first before flowing CO$_2$, and then the atmosphere was switched to flowing 1 atm pure CO$_2$. The temperature was heated up in steps from room temperature. Each XRD 2θ scan was taken after holding the sample for more than 40 minutes at the setting temperature to ensure a complete reaction of powders with CO$_2$. The same scan process was repeated up to 900 °C or 1000 °C. The experimental process of temperature vs. time for the *in-situ* XRD is given in Fig. 1.

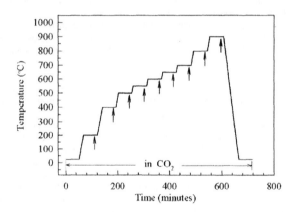

Fig. 1 The experimental process (temperature vs. time) of in-situ temperature-dependent x-ray diffraction in 1 atm pure CO$_2$ atmosphere. The arrows indicate the positions to start the XRD scan.

RESULTS AND DISCUSSION

In order to identify the chemical decomposition, XRD spectra of BaCO$_3$, CeO$_2$, and Y$_2$O$_3$ powders (99.9% purity) were obtained at room temperature as shown in Figs. 2(a)–2(c). The strongest 2θ peaks of BaCO$_3$, CeO$_2$, and Y$_2$O$_3$ appears at 24.2°, 28.5° and 29.2°, respectively.

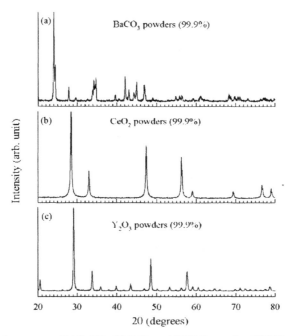

Fig. 2 X-ray diffraction spectra of (a) BaCO$_3$, (b) CeO$_2$, and (b) Y$_2$O$_3$ powders (99.9%) measured in air at room temperature.

Figure 3(a) shows the XRD spectrum of BZCY442 powder before exposure to 1 atm pure CO$_2$–flowing atmosphere. The main 2θ–diffraction peaks include (110), (200), (211), (220), (310), and (222), suggesting a body-centered cubic (bcc) unit cell according to the structure-factor calculation.[16] The cubic lattice parameter was estimated from the (110) peak (2θ=29.2°) and is a=4.322 Å. Note that two weak peaks of second phase were observed between (111) and (200) peaks and near the (310) peak as indicated by "*". In 1 atm pure CO$_2$ atmosphere upon heating, the XRD spectra remain the same below 550 °C without obvious chemical decomposition as shown in Figs. 3(b)–3(d). This is consistent with the earlier result from Ba(Zr$_{0.1}$Ce$_{0.7}$Y$_{0.2}$)O$_{3-\delta}$ (BZCY172),[12] which exhibited a chemical stability in 2% CO$_2$–containing atmosphere at 500 °C for one week. However, at 600 °C (Fig. 3f) an obvious peak appears at 2θ=23.5°, indicating appearance of BaCO$_3$ as denoted by "A". In addition, four weak peaks at least were observed at about 28.5°, 33.5°, 47.9°, and 56.5° as indicated by "B", corresponding to the cubic Ce(Zr,Y)O$_2$ according to the XRD spectrum of CeO$_2$ (Fig. 2b). As temperature increases, the relative intensities of "A", "B", and "C" peaks grow gradually up to 800 °C. The "C" peak indicates the Y$_2$O$_3$–like structure, which begin to appear visibly at 700 °C as seen in Fig. 3(g). At 900 °C (Fig. 3j) "A" peak of BaCO$_3$ disappears and two new peaks appear at 21.4° and 25.9° as indicated by "D"s, confirming a transformation from orthorhombic to hexagonal structures.[17,18]

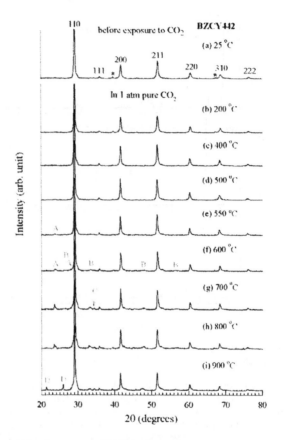

Fig. 3 *In-situ* x-ray diffraction spectra of BZCY442 upon heating. A, B, and C peaks correspond to BaCO$_3$, Ce(Zr,Y)O$_2$, and Y$_2$O$_3$-like structure, respectively. "D" indicates a formation of hexagonal structure. The same notations are also used in Figs. 4 and 5.

Figure 4(a) shows the XRD spectra of BZCY532 powder before exposure to CO$_2$–containing atmosphere. The cubic lattice parameter was estimated from the Cu Kα_1 (110) peak ($2\theta=29.3°$) and is a=4.307 Å, which is slightly less than a=4.322 Å of BZCY442 due to the smaller ionic radius of Zr^{4+} (R^{IV}=0.72 Å) relative to that of Ce^{4+} (R^{IV}=0.87 Å).[7] As was observed also in BZCY442, the XRD spectra of BZCY532 powders remain unchanged up to 550 °C as seen in Figs. 4(b)–(e). At 600 °C, BaCO$_3$, Ce(Zr,Y)O$_2$, and Y$_2$O$_3$-like structures appear noticeably as indicated by "A", "B", and "C", respectively. As temperature increases, the relative intensities of "A", "B", and "C" peaks grow gradually up to 800 °C. Above 800 °C as seen in Figs. 4(j) and 4(k), a formation of hexagonal structure takes place, associated with two new peaks at 21.3° and 25.9° as indicated by "D". Note that peak "C" disappears at 1000 °C as seen in Fig. 4(j).

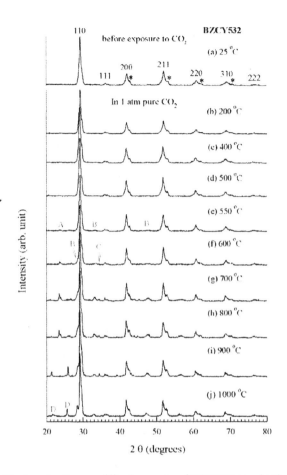

Fig. 4   *In-situ* x-ray diffraction spectra of BZCY532 upon heating.

The XRD spectrum of BZCY622 powders before exposure to 1 atm pure CO_2 atmosphere is given in Fig. 5(a). The cubic lattice parameter was estimated from the (110) peak ($2\theta=29.5°$) and is $a=4.279$ Å, which is less than those of BZCY532 and BZCY442. A weak second-phase peak was observed between (111) and (200) peaks as indicated by "*". As evidenced in Figs. 6(b)–6(i), BZCY622 remains stable up to 900 °C without an obvious $BaCO_3$ peak, indicating that BZCY622 is chemically stable in 1 atm pure CO_2 atmosphere.

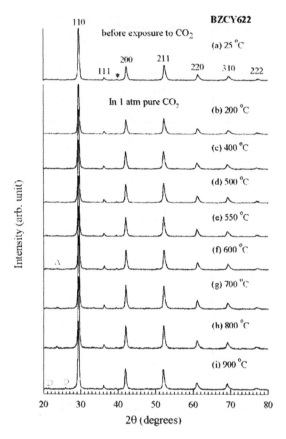

Fig. 5 *In-situ* x-ray diffraction spectra of BZCY622 upon heating.

CONCLUSIONS

Ba(Zr$_{0.8-x}$Ce$_x$Y$_{0.2}$)O$_{3-\delta}$ (x=0.2, 0.3, and 0.4) exhibit a sufficient chemical stability below 550 °C in 1 atm pure CO$_2$–flowing atmosphere without obvious decomposition to BaCO$_3$ and Ce(Zr,Y)O$_2$. As temperature increases, the low-cerium compounds Ba(Zr$_{0.8-x}$Ce$_x$Y$_{0.2}$)O$_{3-\delta}$ ($x\leq0.2$) remain stable in a CO$_2$ environment at least up to 900 °C. BaCO$_3$ proceeds a first-order structure transformation of orthorhombic–hexagonal in the temperature region of 800–900 °C in BZCY442 and BZCY532 upon heating. This study suggests that low-cerium ceramics Ba(Zr$_{0.8-x}$Ce$_x$Y$_{0.2}$)O$_{3-\delta}$ ($x\leq0.2$) are promising candidates for applications of intermediate–temperature proton-conducting SOFCs. Note that these tests in this work were done in 1 atm pure CO$_2$ atmosphere. One can expect that Ba(Zr$_{0.8-x}$Ce$_x$Y$_{0.2}$)O$_{3-\delta}$ ceramics may exhibit even better stability if the CO$_2$ concentration is lower.

This work was supported by National Science Council of Taiwan Grant No. 96-2112-M-030-001, and DOE under subcontract DE-AC06-76RL01839 from Battelle Memorial Institute and PNNL.

REFERENCES
1. J.P. Trembly, R.S. Gemmen, and D.J. Bayless, J. Power Sources 163 (2007) 986-996.
2. R.S. Gemmen and J. Trembly, J. Power Sources 161 (2006) 1084-1095.
3. S. McIntosh and R.J. Gorte, Chem. Rev. 104 (2004) 4845-4865.
4. S.M. Haile, Acta Materialia 51 (2003) 5981-6000.
5. C.W. Tanner and A.V. Virkar, J. Electrochem. Soc. 143 (1996) 1386-1389.
6. S.V. Bhide and A.V. Virkar, J. Electrochem. Soc. 146 (1999) 2038-2044.
7. K. H. Ryu and S. M. Haile, Solid State Ionics 125 (1999) 355-367.
8. G. Ma, T. Shimura, and H. Iwahara, Solid State Ionics 110 (1998) 103-110.
9. K. Katahira, Y. Kohchi, T. Shimura, and H. Iwahara, Solid Sate Ionics 138 (2000) 91-98.
10. F.M.M. Snijkers, A. Buekenhoudt, J. Cooymans, and J.J. Luyten, Scripta Materialia 50 (2004) 655-659.
11. A. Magrez and T. Schober, Solid State Ionics 175 (2004) 585-588.
12. C. Zuo, S. Zha, M. Liu, M. Hatano, and M. Uchiyama, Advanced Materials 18 (2006) 3318-3320.
13. A.S. Patnaik and A.V. Virkar, J. Electrochem. Soc. 153 (2006) A1397-A1405.
14. P. Babilo, T. Uda, and S.M. Haile, J. Mater. Res. 22 (2007) 1322-1330.
15. P. Babilo and S.M. Haile, J. Am. Ceram. Soc. 88 (2005) 2362-2368.
16. B.D. Cullity, Elements of X-ray Diffraction (Addison-Wesley Publishing, 1978), 2nd edition.
17. R. Strobel, M. Maciejewski, S.E. Pratsinis, A. Baiker, Thermochimica Acta 445 (2006) 23-26.
18. C.R.M. Rao, P.N. Mehrotra, J. Thermal Analysis 17 (1979) 539-542.

EVALUATION OF THE RESIDUAL STRESS PROFILES OF PRACTICAL SIZE LANTHANUM GALLATE-BASED CELLS IN RADIAL DIRECTION

Hiroyuki Yoshida, Mitsunobu Kawano, Koji Hashino, Toru Inagaki, Hiroshi Deguchi, and Yoshiyuki Kubota
The Kansai Electric Power Company, Inc.
11-20 Nakoji 3-chome, Amagasaki
Hyogo 661-0974, Japan

Kei Hosoi
Mitsubishi Materials Corporation
1002-14 Mukohyama, Naka-shi
Ibaraki 311-0102, Japan

ABSTRACT
        The residual stress profiles in radial direction of commercial size cells (120mm diameter) with $La_{0.8}Sr_{0.2}Co_{0.8}Mg_{0.15}Co_{0.05}O_{3-\delta}$ (LSGMC) electrolyte were measured by X-ray diffraction techniques using synchrotron radiation. A composite of NiO, Ru, and $Ce_{0.8}Sm_{0.2}O_{1.9}$ (SDC) was used as anode and $Sm_{0.5}Sr_{0.5}CoO_{3-\delta}$ (SSC) was used as cathode. The residual stress in each anode and cathode component was almost constant in the radial direction except at the edges at a penetration depth of 15 μm. The change of the stress at the edges is considered to occur during the manufacturing of the electrolyte. The compressive stresses in the electrolyte at the cathode side were in the sequence of LSGMCc < LSGMCa, and the tensile stresses of the electrode components were in the sequence of Anode (average value of Ni and SDC) < SSC. The relationship of the size order of the residual stresses shows the residual stresses in the cells tend to warp cathode to inside. The residual stress differences between anode and cathode interfaces showed that the values decreased from the center to the edge. The total residual stress at the edge of the cell operated at high power density was much smaller than that at the center. That cell has a large warp, therefore, the behavior of the total residual stresses is considered to increase the stress at out-of-plane direction instead of decreasing the stress at in-plane direction. It can be said that the presence of the residual stress at out-of-plane direction can be estimated by examining the variation of total residual stresses from the center part to the edge part.

INTRODUCTION
        In recent years, intermediate temperature (600-800°C) solid oxide fuel cells (IT-SOFCs) have been drawing much attention because of the possibility of using internal reforming of hydrocarbon fuel and metallic interconnects. It has been reported that doped lanthanum gallate-based compounds exhibit high ionic conductivity at intermediate temperatures and are attractive as electrolyte in this type of cells[1]. The Kansai Electric Power Company Inc. (KEPCO) and Mitsubishi Materials Corporation (MMC) have been developing IT-SOFCs with lanthanum gallate based electrolyte (La,Sr)(Ga,Mg,Co)$O_{3-\delta}$ (LSGMC), Ni-(Ce,Sm)$O_{2-\delta}$ anode (Ni-SDC) and (Sm,Sr)$CoO_{3-\delta}$ cathode (SSC). Ni-SDC cermet anode gives high performance due to synergistic effect of enlarging reaction area and increasing paths for ionic and electronic conduction. This kind of anode is originally based on KEPCO's technology[2-4]. LSGMC electrolyte and SSC cathode are based on MMC's technologies collaborated with Ishihara's group of Kyushu University[5,6]. This cell of 120 mm in diameter showed good performance. Typical output power density obtained using a single cell is 0.25 W/cm^2 with fuel utilization of 70% at 750°C[7]. Utilizing these cells in a 1kW-class module, maximum DC efficiency of 54%HHV was obtained, and a 10kW-class CHP system equipped with these cells has been successfully operated[8].

KEPCO and MMC have been trying to increase electric power output per single cell to obtain higher volumetric power density and lower cost, which will reduce the quantity of module components and the cell-stack size. The module should be operated at higher current density to obtain higher power density, and we have been working to achieve further increase in power density under a NEDO (New Energy and Industrial Technology Development Organization) supported project.

High power density operation generates large amount of heat and water vapor. It may cause cell crack and/or delamination by increasing of the internal stresses in the cells due to heat and atmosphere distribution. Thus, it is necessary to evaluate the residual stresses in the developed cell for high power density. The evaluation of residual stresses of the materials in lanthanum gallate based cells was examined by X-ray diffraction (XRD) with synchrotron radiation [9]. The depth profiles of the residual stresses in anode and cathode of the standard cell were constant within the range of 5 - 30 μm, as was shown in the previous study [9]. However, the previous study was examined in the depth direction at the center of the cell, and the residual stress distribution in the radial direction was not clarified. The developed cell stack has the seal-less structure, and it is assumed that the stress distribution is generated during high power density operation because the atmosphere and the temperature at the center of the cell is considered to be different from those at the edge of the cell. Therefore, the residual stress profiles in radial direction of commercial size cells with LSGMC electrolyte were measured by X-ray diffraction.

EXPERIMENTAL

Sample Preparation

Cells composed of $La_{0.8}Sr_{0.2}Ga_{0.8}Mg_{0.15}Co_{0.05}O_{3-\delta}$ (LSGMC) electrolyte, $Sm_{0.5}Sr_{0.5}CoO_{3-\delta}$ (SSC) cathode, and $Ni-Ru-Ce_{0.8}Sm_{0.2}O_{2-\delta}$ (SDC) cermet anode were fabricated. The current standard cell has a diameter of 120 mm and electrolyte thickness of 200 μm. The source powder for the electrolyte is prepared by conventional solid state reaction techniques using commercially available powders of La_2O_3, $SrCO_3$, Ga_2O_3, MgO and Co_3O_4. The mixed powders are ball-milled and calcined in air. The calcined powder is re-ground and mixed with a solvent and an organic binder to be tape-cast into a green sheet. The green sheet was cut and sintered in air at 1400°C for 6 h. The relative density of LSGMC was greater than 98%.

For the anode, the slurry composed of a mixture of NiO and SDC was screen-printed onto the electrolyte and then sintered in air at 1200°C for 3 h. After anode sintering, the slurry made of SSC powder was screen-printed onto the electrolyte and sintered at 1100°C for 3 h. Thickness values of porous electrode layers are in the range between 30 and 50 μm. In order to accelerate reforming and electrochemical reaction of hydrocarbon fuel at anode, Ru nanoparticles were dispersed on the anode. After cathode sintering, colloid solution with Ru nanoparticles was dropped onto NiO-SDC anode and dried at 120°C.

Table I summarizes the components of the cells measured in this study. The cells of 120 mm in diameter were used for the measurement without dividing them into pieces. The sample with NiO-SDC anode and SSC cathode sintered onto the electrolyte followed by infiltration of Ru-containing solution and drying was described as Cell 1. Cell 2 was made by dual-atmosphere heat-treatment of Cell 1 at 750°C for 1 h, where anode and cathode sides were exposed to hydrogen and air, respectively. After dual-atmosphere heat-treatment, the cell was operated at high power density of 0.4 W/cm^2 (fuel utilization of 75%) for 1 h at 750°C. This cell was described as Cell 3.

Table I. The components of the cells measured in this study.

	Anode	Electrolyte	Cathode	Cell operation
Cell 1	NiO-Ru-SDC	LSGMC	SSC	after drying of Ru-containing colloid
Cell 2	Ni-Ru-SDC	LSGMC	SSC	after anode reduction
Cell 3	Ni-Ru-SDC	LSGMC	SSC	after high power density operation

X-ray Diffraction measurements

The residual stresses were estimated by X-ray diffraction measurement. Synchrotron radiation includes high energy X-ray with high directionality, therefore, sufficient diffraction intensity can be detected from small radiation area of 1 * 1 mm. High-energy X-ray is included in a synchrotron radiation. The residual stress measurement inside the component becomes possible by using high-energy X-ray[10]. Furthermore, the beam is monochromated, and there are not any problems due to the effect of $K_{\alpha1-\alpha2}$ doublet. From these characteristics, we selected a synchrotron radiation as X-ray source.

The residual stresses in the cells were measured at BL16XU line of SPring-8, which was the same line with that measured in the previous study[9]. The typical thickness of the electrode was between 30 and 50 μm, therefore, high-energy X-ray was necessary to measure the diffraction of electrolyte under electrode. Moreover, suitable energy beam should be selected because too high energy beam decreases the measurement precision due to the decreasing of the diffracting angle. Therefore, we selected high-energy X-ray of 37 keV for the stress measurements.

The cell arrangement in XRD system is shown in Fig. 1. X-ray was radiated onto the cells without damage. The cells were fixed with magnets on the stainless steel disk, which was fixed on vacuum chuck. The vacuum chuck was set on slider with a ruler. The beam size at the sample point was adjusted to 1 * 1 mm by four-quadrant silts in front of the sample. The constant penetration depth method was applied for the XRD measurements. In order to maintain the penetration depth constant, a combination of the side-inclination method and the iso-inclination method was used. The sets of the side-inclination angle and iso-inclination angle at each $\sin^2\psi$ were calculated from X-ray absorption coefficient, density, and thickness of each material to keep effective penetration depth constant.

Schematic illustration of X-ray radiation at each cell point is shown in Fig. 2. The penetration depth was fixed at 15 μm from the electrode surface for the measurement of the electrode materials and from the electrode/electrolyte interface for the measurement of the electrolyte material. The residual stress at each cell point in radial direction was measured. The measured points are 0 mm, 13 mm, 27 mm, 41 mm, and 55 mm from the center of the cell. The residual stresses on three points (0 mm, 27 mm, and 55 mm) could not be measured in the case of Cell 1 due to the lack of machine time. Since the position of incident X-ray could not be moved, the samples were slid shown in Fig. 2.

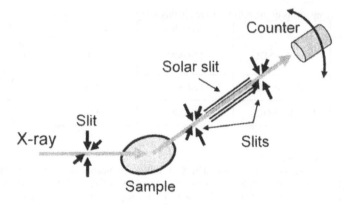

Figure 1.   Schematic illustration of XRD system arrangement.

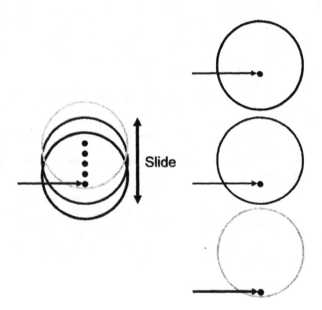

Figure 2.   Schematic illustration of X-ray radiation at each cell point.

Determination of Residual Stresses

The residual stresses were estimated by the $\sin^2\psi$ method[11,12]. Details about the method were described in a previous paper[9], and the summary was described below.

The relationship between 2θ and $\sin^2\psi$ is described in Eq. 1.

$$2\theta = -\frac{2(1+v)}{E}\sigma\tan\theta_0\sin^2\psi + \frac{4v}{E}\tan\theta_0\sigma + 2\theta_0 \qquad (1)$$

The residual stress in the sample, inclination angle, and the X-ray elastic constants are σ, ψ, $(1+v)/E$, respectively. Bragg angles without distortion and measured angle are θ_0 and θ, respectively. The residual stress σ was determined from the slope of 2θ - $\sin^2\psi$ plots. Before the measurement of the cell samples, the 2θ - $\sin^2\psi$ plots of the CeO_2 powder as the standard sample was taken, and the change in the peak position according to an increase in $\sin^2\psi$ was approximated in the straight line. The value of the slope obtained from the 2θ - $\sin^2\psi$ plots of the cell sample was corrected by the value of the slope obtained for the measurement of the CeO_2 powder.

X-ray Elastic Constant Measurement

In order to determine the residual stress from the slope of Eq. 1, the X-ray elastic constants should be derived separately. The X-ray elastic constants of the components were measured using the laboratory XRD apparatus with Cu X-ray tube (Rigaku, PSPC/MSF-2M). The 2θ range is limited between 140 and 160° owing to the property of this apparatus. Furthermore, the diffracting planes of the components must be separated from other peaks. Thus, we selected the diffracting planes of the components as shown in Table II. In the case of the electrolyte, X-ray beam was radiated to both sides, that is, the residual stresses at anode and cathode sides were derived. XRD measurements of ruthenium compounds were not examined since the concentration of added Ru was too low to detect the peaks of the diffracting planes.

Table II. The diffracting planes of the components measured in this study.

Material	Component	Diffracting plane	$2\theta_0$ (° at 37 keV)
Ni	Anode	331	23.95
NiO	Anode	511/333	24.05
SDC	Anode	622	23.59
LSGMC	Electrolyte	422	24.29
SSC	Cathode	332	23.83

RESULTS

The peak positions of the diffracting planes for electrode and electrolyte materials were derived from Gaussian fitting of the XRD patterns. Fig. 3 shows 2θ - $\sin^2\psi$ plots of cathode side components of Cell 1. Left and right figures are the plots of SSC and LSGMCc, respectively. LSGMCc means the residual stress of LSGMC measured by radiating X-ray from cathode side. The numbers in legends mean the distance from the center of the cell in radial direction. In the case of SSC, the peak position shifted to lower angle with increasing of $\sin^2\psi$, which showed that the tensile stress resided in SSC. Furthermore, the peak positions of all measured points at each $\sin^2\psi$ were almost the same value, which indicated that almost the same residual stress existed in any points. On the other hand, it was found from the variation of LSGMCc peaks that the compressive stress resided in LSGMCc since the peak position shifted to higher angle with increasing of $\sin^2\psi$. In the case of LSGMCc, the variation of the peaks at 0 mm was almost the same as that at 27 mm. However, the variation of the peaks at 55 mm was different from those at other points; therefore, the difference stress was expected to reside at around the edge of the electrolyte. In this way, 2θ - $\sin^2\psi$ plots of each element of all samples were examined, and the residual stresses were calculated from the slope of the fitted lines.

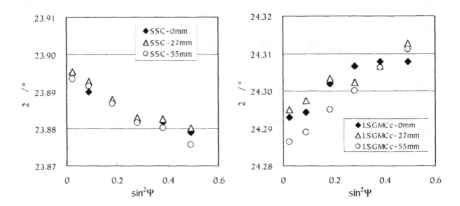

Figure 3. 2θ - $\sin^2\psi$ plots of cathode components of Cell 1: left and right figures are those of SSC and LSGMCc, respectively.

Fig. 4 shows the residual stress profiles of the components of the cells in the radial direction. LSGMCa means the residual stress in the electrolyte measured with radiation from anode side. The positive and negative values mean tensile and compressive stresses, respectively. The following can be said about all cells. The tensile stresses resided in the components of anode and cathode, and the compressive stresses resided in the part of LSGMC electrolyte beneath anode and cathode, which was consistent with the results of the previous paper[9]. It was found from the data of anode components that the tensile stresses of Cell 2 and Cell 3 (reduced state) were lower than those of Cell 1 (oxidized state), while the residual stress in the cathode was constant. This behavior was also consistent with that reported in the previous paper[9].

The residual stress profiles in the radial direction of electrode components in Cell 1 and 2 were almost constant from the center to the edge. Therefore, it was considered that the electrodes were uniformly printed out and reduced. The residual stress profile in the electrolyte showed some variation. Though the compressive stresses of LSGMCa and LSGMCc at between 0 and 27 mm gave almost the same value, that of LSGMCa at 55 mm was decreased and that of LSGMCc at 55 mm was increased. No warp on Cell 1 was observed. The characteristic behavior of the residual stress at the edge part of the electrolyte, therefore, is considered due to the reflection of the stress that hang to the edge of the green sheet when the electrolyte was cut out from the sheet and/or due to the difference of the in-furnace temperature profile only at the edge part of the electrolyte from the other parts. In the case of Cell 2 at the center, the compressive stresses of LSGMCc and LSGMCa were smaller and larger than those of Cell

1, respectively. The compressive stresses of LSGMCa were constant or slightly decreased from the center part to the edge part. On the other hand, those of LSGMCc were almost constant between 0 and 41 mm, and then drastically increased at 55 mm. The values of compressive stresses in Cell 2 were totally different from those of Cell 1. Furthermore, Cell 2 has some warp; therefore, it was considered that these differences were caused by the influence of anode reduction.

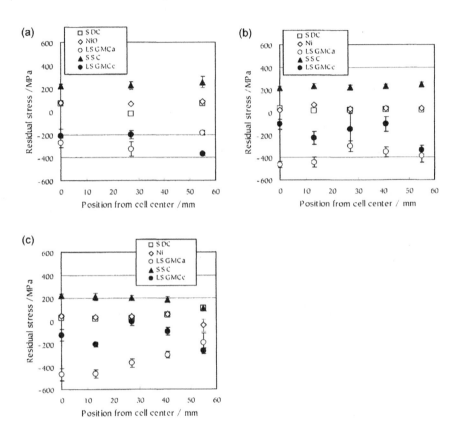

Figure 4. The residual stress profiles of the components of the cells in radial direction; (a) Cell 1. (b) Cell 2, and (c) Cell 3.

Cell 3 is the cell after high power density operation, and the tensile stresses of anode components were decreased by anode reduction as well as those of Cell 2. However, the tensile stress of SDC at the edge part was increased and that of Ni was changed to small compressive stress. The average value of the residual stresses of SDC and Ni at the edge part was almost the same as those at the other parts. This behavior is considered due to the aggregation of Ni and/or the destruction of anode network structure since the atmosphere at the edge part had larger amount of heat and water vapor compared to the center part and air from the outside of the cell. The tensile stress of SSC was also decreased at the edge part, which was assumed to be crystal lattice expansion by reduction due to the higher air utilization result from the high power density operation. The compressive stress of LSGMCc in Cell 3 at the edge part was drastically increased as well as that in Cell 2. The compressive stresses of LSGMCa were decreased from the center part to the edge part. The residual stress profiles of the cell components in radial direction were considered to be related to the large warp in Cell 3.

DISCUSSION

From the results of the residual stress profiles in the radial direction of the cell, the relationship between the residual stresses and the warp in the cell were discussed. There was no warp in Cell 1. The residual stresses of LSGMCa and LSGMCc in Cell 1 were almost the same and constant between 0 and 27 mm and those at the edge part had small difference. This difference did not make a warp since it was occurred during the sintering of the electrolyte. Cell 2 has a small warp, and Cell 3 has a large warp. The directions of the curvature in the cells at anode and cathode sides were outwards and inwards, respectively. The compressive stresses of the electrolyte components were in the sequence of LSGMCc < LSGMCa, and the tensile stresses of the electrode components were in the sequence of Anode (average value of Ni and SDC) < SSC. The relationship of the size order of the residual stresses shows the residual stresses in the cells tend to warp cathode to inside, which is corresponding to the direction of the actual warps. The cells measured in this study were unsealed during operation, therefore, they have not received the stresses from the seals.

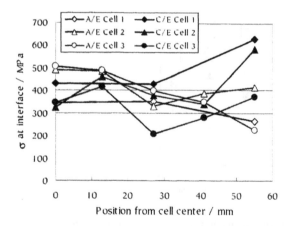

Figure 5. The residual stress profiles in radial direction at anode and cathode interfaces.

For further consideration about the residual stresses in the cells, the residual stresses at the anode/electrolyte interface $\sigma_{(A/E)}$ and at the cathode/electrolyte interface $\sigma_{(C/E)}$ were calculated and plotted in radius direction, as shown in Fig. 5. $\sigma_{(A/E)}$ and $\sigma_{(C/E)}$ are positive values of the residual stress difference between electrode and electrolyte. The residual stress profiles in depth direction in the previous study showed that the residual stresses of the cell materials are constant in the vicinity of the interface[9], therefore, the residual stresses at 15 μm depth from the surface of the components were set to those at the interfaces in calculation. They were between 300 and 500 MPa at the center part, and the range between the largest and the smallest values in the cells except for the edge part were around 200 MPa though the value drifted at each point. On the other hand, the residual stresses in anode interface of the cells at the edge part decreased and those in cathode interface increased. The maximum and minimum residual stresses in the interfaces were about 600 and 200 MPa, respectively. This behavior was evident for all cells. Therefore, it is assumed that the residual stress change at the edge was mainly caused during the electrolyte sintering.

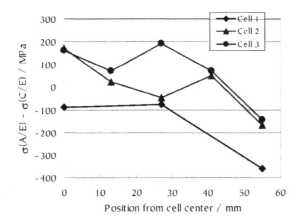

Figure 6. The residual stress differences between anode and cathode interfaces.

In order to clarify the behavior of the residual stress change, the residual stress differences between anode and cathode interfaces in the radial direction were plotted, as shown in Fig. 6. The positive value should mean that the residual stress at anode interface is larger than that at the cathode interface. It was found that the values in all cells decreased from the center towards the edge, which suggests that the larger stress resided in anode side at the center part and the residual stress at cathode interface increased due to relaxation of the residual stress of LSGMCa.

The variation of the sums of $\sigma_{(A/E)}$ and $\sigma_{(C/E)}$ in radial direction was also calculated. The total residual stress at the edge was different from that at the center. In the case of Cell 1, the total residual stress slightly increased at the edge. Though the total residual stresses in Cell 2 were dispersed, it's value at the edge was slightly larger than that at the center. On the other hand, the total residual stress at the edge in Cell 3 was much smaller than that at the center. This is due to the drastic decreasing of $\sigma_{(A/E)}$ and smaller increasing of $\sigma_{(C/E)}$. Cell 3 has a large warp, therefore, the behavior of the total residual stresses is considered to increase the stress at out-of-plane direction instead of decreasing the stress at in-plane direction. The analysis method in this study is based on Eq. 1, which is approximated disregarding stress at out-of-plane direction. Therefore, it is necessary to be careful about the interpretation of the result when measuring a sample having a large warp. That is, the decrease of the total residual stress at in-plane direction by high power density operation does not mean the distribution of the residual stress was relaxed and/or averaged. This behavior means the residual stress generated at out-of-plane direction, and it may be a cause of the cell destruction. It can be said from these results that the presence of the residual stress at out-of-plane direction can be estimated by examining the variation of total residual stresses from the center part to the edge part.

CONCLUSIONS

The residual stress profiles of practical size cells in radial direction were measured by XRD. The residual stress at the edge part of each component of the cell decreased after an electric power generation. The tensile stresses resided in the components of anode and cathode, and the compressive stresses resided in the part of LSGMC electrolyte beneath anode and cathode. The residual stress profile showed that the stress at the edge part was different from those at the other parts. This behavior is considered to occur during the manufacturing of the electrolyte. The compressive stresses of the electrolyte components were in the sequence of LSGMCc < LSGMCa, and the tensile stresses of the electrode components were in the sequence of Anode (average value of Ni and SDC) < SSC. The magnitude of the residual stresses for anode and cathode side shows the residual stresses in the cells tend to warp the cathode inwards, which corresponds to the direction of the actual warps. The residual stresses at both the anode/electrolyte interface and the cathode/electrolyte interface were calculated. The range between the largest and the smallest values in the cells except for the edge part were around 200 MPa, and those at the edge part were spread to about 400 MPa. The residual stress differences between anode and cathode interfaces in radial direction were calculated, and it was found that the values in all cells decreased from the center to the edge, which indicated that the larger stress resided in the anode side at the center part and the residual stress at cathode interface increased due to relaxation of the residual stress of LSGMCa. The total residual stress at the edge in the cell operated at high power density was much smaller than that at the center. That cell has a large warp, therefore, the behavior of the total residual stresses is considered to increase the stress at out-of-plane direction instead of decreasing the stress at in-plane direction. It can be concluded from these results that the presence of the residual stress at out-of-plane direction can be estimated by examining the variation of total residual stresses from the center part to the edge part.

ACKNOWLEDGMENTS

The authors would like to express their gratitude to New Energy and Industrial Technology Development Organization for supporting the SOFC development project.

The synchrotron radiation experiments were performed at SPring-8 with the approval of the Japan Synchrotron Radiation Research Institute (JASRI) (Proposal No. 2006B5050).

The authors are grateful to Prof. Keisuke Tanaka of Nagoya University for his advice about XRD measurement.

REFERENCES
[1]T. Ishihara, H. Matsuda, and Y. Takita, Doped LaGaO Perovskite Type Oxide as a New Oxide Ionic Conductor, *J. Am. Chem. Soc.*, **116**, 3801-3 (1994).
[2]S. Ohara, R. Maric, X Zhang, K. Mukai, T. Fukui, H. Yoshida, T. Inagaki, and K. Miura, High performance electrodes for reduced temperature solid oxide fuel cells with doped lanthanum gallate electrolyte: I. Ni–SDC cermet anode, *J. Power Sources*, **86**, 455-8 (2000).
[3]H. Yoshida, T. Inagaki, K. Hashino, M. Kawano, H. Ijichi, S. Takahashi, S. Suda, and K. Kawahara, Study on the Network-Formation Process in the Sintering of the NiO-SDC Composite Powders Prepared by Spray Pyrolysis Method, *in Solid Oxide Fuel Cells IX*, S.C. Singhal and J. Mizusaki, Editors, **PV 2005-07**, The Electrochemical Society Proceedings Series, Pennington, PJ, 1382-89 (2005).

[4]M. Kawano, K. Hashino, H. Yoshida, H. Ijichi, S. Takahashi, S. Suda, and T. Inagaki, Synthesis and characterizations of composite particles for solid oxide fuel cell anodes by spray pyrolysis and intermediate temperature cell performance. *J. Power Sources*, **152**, 196-9 (2005).

[5]T. Ishihara, M. Honda, T. Shibayama, H. Minami, H. Nishiguchi, and Y. Takita, Intermediate Temperature Solid Oxide Fuel Cells Using a New LaGaO₃ Based Oxide Ion Conductor, *J. Electrochem. Soc.*, **145**, 3177-83 (1998).

[6]K. Kuroda, I. Hashimoto, K. Adachi, J. Akikusa, Y. Tamou, N. Komada, T. Ishihara, and Y. Takita, Characterization of solid oxide fuel cell using doped lanthanum gallate, *Solid State Ionics*, **132**, 199-208 (2000).

[7]M. Kawasaki, N. Chitose, J. Akikusa, T. Akbay, H. Eto, T. Inagaki, and T. Ishihara, Lanthanum Doped Barium Cobaltite as a Novel Cathode for Intermediate-Temperature SOFC Using Lanthanum Gallate Electrolyte, *ECS Transactions*, **7**, 1229-34 (2007).

[8]M. Shibata, N. Murakami, T. Akbay, H. Eto, K. Hosoi, H. Nakajima, J. Kano, F. Nishiwaki, T. Inagaki, and S. Yamasaki, Development of Intermediate-Temperature SOFC Modules and Systems, *ECS Transactions*, **7**, 77-83 (2007).

[9]H. Yoshida, H. Deguchi, T. Inagaki, K. Hashino, M. Kawano, K. Hosoi, and M. Horiuchi, Residual Stress Analysis in Lanthanum Gallate-Based Cell before and after Fuel Cell Operation, *ECS Transactions*, **7**, 429-34 (2007).

[10]P.J. Withers, M. Preuss, P.J. Webster, D.J. Hughes, and A.M. Korsunsky☐Residual strain measurement by synchrotron diffraction, *Mater. Sci. Forum*, **404-407**, 1-12 (2002).

[11]Y. Yoshioka, K. Hasegawa, and K. Mochiki, Stress measurement in Stainless Steel by Use of Monochromatic Cr-K beta X-Rays and a Position Sensitive Detector, *Adv. X-Ray Anal.*, **24**, 167 (1981).

[12]M. Barral, J.M. Sprauel, J. Lebrum, G. Maeder, and S. Megtert, On the Use of Synchrotron Radiation for the Study of the Mechanical Behaviour of Materials, *Adv. X-Ray Anal.*, **27**, 149-158 (1984).

Electrodes

EFFECT OF SPRAY PARAMETERS ON THE MICROSTRUCTURE OF $La_{1-x}Sr_xMnO_3$ CATHODE PREPARED BY SPRAY PYROLYSIS

Hoda Amani Hamedani, Klaus-Hermann Dahmen, Dongsheng Li, Hamid Garmestani
School of Materials Science and Engineering, Georgia Institute of Technology,
Atlanta, GA 30332, USA

ABSTRACT

Manufacturing high-performance cathodes requires optimization of conventional processing techniques to novel ones capable of controlling the microstructure. Spray pyrolysis is one of those promising techniques for tailoring microstructure of the electrodes for better performance of solid oxide fuel cells (SOFCs). This paper reports the effect of solvent and precursor type, deposition temperature and spray speed on morphology and compositional homogeneity of the lanthanum strontium manganite (LSM) cathode. Results show that metal-organic precursors and organic solvent create a homogeneous crack-free deposition as opposed to aqueous solution. By changing the temperature gradually from 540 to 580 °C and spray speed from 0.73 to 1.58 ml/min, an appreciable trend was observed in amount of porosity in LSM cathode microstructure. It was shown that increasing the temperature and spray speed results in formation of more porous microstructure. The microstructure, morphology and the compositional homogeneity of the fabricated cathodes were characterized using SEM, EDS and XRD.

INTRODUCTION

Solid oxide fuel cells (SOFCs) with a high chemical to electrical energy conversion efficiency attracted considerable interest during the last decades. Large scale production of thin film electrodes and development of electrodes with high electrocatalytic activity for long-life intermediate temperature SOFCs require introducing novel manufacturing methods. During recent years, many research efforts have been focused to design materials able to provide high oxygen permeation rates and long term stability in real applications. However, few studies have been conducted to introduce novel techniques capable of offering optimized properties based on microstructure design. Lanthanum strontium manganate (LSM) is widely used as a cathode material due to its high electrical conductivity, relatively high electrocatalytic activity for O_2 reduction, and good thermal and chemical compatibility with yttria stabilized zirconia (YSZ) electrolyte[1-5].

The cathode microstructure has been recognized as one of the principle factors determining the performance of the SOFC, since the electrical and chemical properties and thermal stability of SOFC depend on the chemical composition and morphology of the electrode. High mixed conductivity, chemical and thermal expansion compatibility, better resistance to creep and failure, and high specific surface area are among those critical properties of cathode electrodes that need to be obtained at the same time under working conditions. Recently, microstructure sensitive design (MSD) has been applied to propose a multi-scale model based on statistical continuum mechanics for the cathode electrode with optimized electrical and mechanical properties.[6-8] According to this model, a gradient porous microstructure with a fine microstructure layer close to the surface of the electrolyte and a coarse outer layer is expected to show much more compatibility of properties in different layers. The highest activity is achieved when the outer layer consists of high porosity that allows more oxygen molecules to reach larger active reaction sites at the inner less porous layers. As the particles get smaller, a larger number of TPB (as the three-dimensional sites between LSM particles, YSZ electrolyte and gas-phase O_2) along with a higher porosity can be developed so that the reaction rate dramatically increases.[9]

Various physical and chemical processes have been used for manufacturing porous LSM electrode such as tape casting, sol-gel[10] gel-casting[11] and screen printing which suffer from crack formation.[12-14] Spray pyrolysis (SP), as one of the most cost effective techniques for deposition of uniform thin films of large surface areas under atmospheric conditions, is performed at a relatively low temperature and at a high deposition rate. SP has the capability of controlling the shape, size, composition and phase homogeneity of the particles by changing spray parameters. The extremely large choice of precursors along with simple equipments for mass production of large areas have made spray pyrolysis an industry compatible manufacturing technique for synthesis of porous films with well-controlled microstructures.[15-17]

In this paper, the effect of spray parameters on the microstructure of LSM cathode is investigated. Also, SP is developed to produce a gradient porous LSM cathode layer with the processing conditions optimized through varying the deposition parameters.

EXPERIMENTAL

A. Preparation of the YSZ electrolyte substrates

YSZ electrolyte substrates were prepared from pressing 8 mol% yittria-stabilized zirconia (TZ8Y) powders from Tosoh Co. into pellets of 2.5 cm in diameter and thickness of about 1 mm by a hydrostatic press (Buehler Ltd). To make a dense electrolyte substrate, the pellets were sintered at the heating rate of 4 °C/min up to 1400 °C at air atmosphere, held for 30 minutes and then cooled down to room temperature.

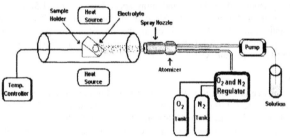

Figure 1. Schematic of the spray pyrolysis set-up.

B. Deposition of porous LSM cathode

Figure 1 shows the schematic of spray pyrolysis set-up consisting of a 1.6 MHz atomizer, two quartz heating bulbs and a horizontal tubular flow reactor. The experiment was conducted in two sections. First, two starting systems were examined; the starting aqueous solution was made by dissolving stoichiometric ratios of metal acetates and metal nitrates precursors of lanthanum, strontium and manganese in deionized water and was sprayed to the YSZ substrate at 480°C. Also, a pure organic solution containing metal-organic precursors was made and sprayed on the YSZ at the same temperature and spray speed as for aqueous solution. The 2,2,6,6-tetramethyl-3,5 heptanedionates-TMHD (Strem Co) as the metal-organic precursors of lanthanum, strontium, and manganese (also known as La(TMHD)$_3$, Sr(TMHD)$_2$, Mn(TMHD)$_3$) were dissolved in appropriate ratio into ethylene glycol dimethyl ether as the organic solvent to achieve reproducible stoichiometric ratio of the film composition La:Sr:Mn = 0.75:0.25:1.

Next, the same stoichiometric ratios of metal-organic precursors were dissolved in ethylene glycol dimethyl ether as organic solvent. To determine the effect of temperature and spray speed on the LSM microstructure, one variable was kept constant while increasing the other one. The solutions were deposited on the YSZ substrate at three different temperatures of 540, 560 and 580°C and spray speeds of 0.73, 1.13, 1.58 ml/min(Table I). All samples were further heat treated to 700°C for 4 hours. The nozzle to substrate distance was kept constant at 6.4 cm in all stages of experiment. The average spray time for all samples was around 30-45 min depending on the spray conditions.

Table I. Spray Conditions for the Experimental Stages.

Stage	Spray Condition / Solvent type	Precursors (mg)			Solvent (ml)	Temperature (°C)	Spray speed (ml/min)	Pyrolysis process	LSM film composition	YSZ electrolyte
A	Aqueous	Lanthanum acetate	Strontium acetate	Manganese acetate	Water	480	0.73	Endothermic (vaporization)	Inhomogeneous	Crack growth/YSZ failure
		110.6	30	86	25					
	Organic	Tris Lanthanum	Bis Strontium	Tris Manganese	Ethylene glycol dimethyl ether	480	0.73	Exothermic	Homogeneous	No cracks
		80.3	22.7	100.7	25					
B	Organic	80.3	22.7	100.7	25	540, 560, 580	0.73,1.13,1.58	Exothermic	Homogeneous	No cracks

Crystalline phases of as-prepared and heat treated films in each stage were characterized by X-ray diffraction using Philips X'Pert MRD Diffractometer, performed at 40 kV and 45 mA using Cu-Kα monochromatic radiation. The particle morphology, pores microstructure and compositional homogeneity of the film were studied by SEM (scanning electron microscopy) and EDX (energy dispersive analysis of X-ray).

RESULTS AND DISCUSSION

A. Effect of precursor materials and the solvent type on the microstructure of LSM

The XRD spectra of samples produced in stage A (Table I) from both aqueous and organic solvents before and after heat treatment are provided in Figure 2(a) and (b), respectively. The X-ray patterns confirmed the perovskite structure of the LSM before and after heat treatment. The peaks show that the LSM is crystalline even at lower temperatures before heat treatment. Also, sharp peaks of crystal phases imply increase of crystallinity after additional heat treatment, preferably, in <121> direction in both samples.

As the SEM micrographs and cross section images of both samples in Figure 3 and Figure 4 indicate, solvent type contributes to the quality and morphology of the deposited film[18]. The LSM film made of aqueous solution shows extensive cracks through out the surface. As a result of an endothermic vaporization of water in aqueous solution the deposition surface gets much colder than the heated surface of the YSZ. Therefore, temperature gradient generates stress and strain distribution in the pressed YSZ pellets result in growth of immature cracks in the LSM film and cause final failure of the YSZ substrate (Figure 3(a) and (b)). Moreover, layers of LSM on YSZ processed in water as the solvent show compositions that are not always very homogenous throughout the sample.

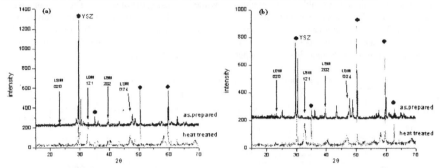

Figure 2. The XRD spectra of samples produced in stage A from (a) aqueous and (b) organic solvents before and after heat treatment.

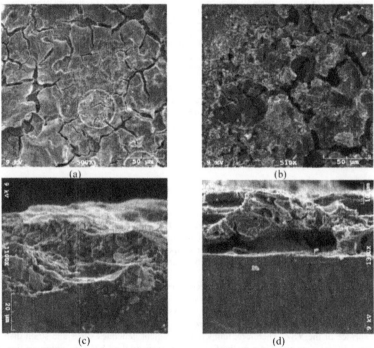

Figure 3. (a), (b) SEM micrographs and (b), (d) cross sections of the sample made from aqueous solution; before heat treatment (left) and after heat treatment (right).

As the cross sections in Figure 3(c) and (d) show, there are more pores and channels in the aqueous sample after heat treatment. However, in organic route, YSZ can act as an oxidizing catalyst helping the pyrolysis of metal-organic precursors and organic solvent. This pyrolysis process is highly exothermic and therefore the temperature difference between sprayed film and the heated YSZ surface is minimal. Therefore, LSM film and YSZ substrate remained without cracking in organic route as shown in Figure 4(a). However, formation of microcracks in the heat treated sample (Figure 4(b)) might be related to the fast cooling of the sample. Examination of the cross sections in Figure 4(c) and (d) reveals that the microstructure through the thickness did not show a lot of changes before and after heat treatment. Both samples show slightly a porous microstructure which could be related to the very low temperature of deposition. Results of this stage are summarized in Table I. Organic route based on metal-organic precursors and organic solvent with lower decomposition temperatures seems to be more favorable to improve morphology and compositional homogeneity. In fact, the organic route has the advantage of easier controlling of the size or even the shape of the particles through changing the spray parameters.

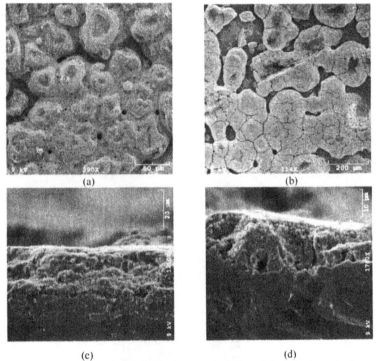

(a)

(b)

(c)

(d)

Figure 4. (a), (b) SEM micrographs and (b), (d) cross sections of the sample made from organic solution; before heat treatment (left) and after heat treatment (right).

B. Investigation of essential parameters to control the microstructure of LSM

a. The effect of temperature

Results verify that the substrate temperature is the most effective parameter to control the morphology of the deposited film.[15] Figure 5 shows a comprehensive comparison of the SEM cross sections of the samples in terms of deposition temperatures and spray speed. The first row of the image shows formation of dense layers at 540°C. As the temperature increases to 560°C. porous microstructure is gradually appeared. According to Figure 5, the most significant porous microstructures formed at different three spray speeds are observed to be related to higher temperatures. In fact. higher deposition temperatures help to decompose the precursors and result in formation of highly porous film. However. higher temperatures over 700°C don't produce any reasonable deposition due to "gas phase nucleation" which usually produces nano-powders that will not deposit on the YSZ and may be carried away by the carrier gas.

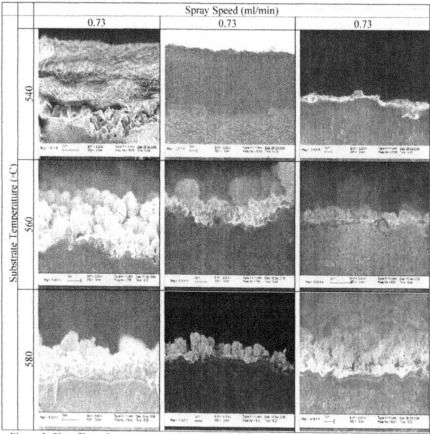

Figure 5. The effect of temperature and the spray speed on the microstructure of the LSM cathode film deposited on the YSZ electrolyte

At the low end of the high temperature range we find smaller particle formations that are embedded in the layer produced by the spray process, leading to a more porous structure (Figure 5). This process is even more pronounced if the speed is increased.

b. The effect of spray speed
    As Figure 5 shows, the film deposited using the lower spray speed looks slightly denser than those deposited at higher spray speed, particularly, at low temperature. Increasing speed leads to much thinner layers at lower temperatures, since it does not give enough activation energy and time to decompose the precursors. Therefore the spray and vapors are transported by the carrier gases away from the YSZ. However, increasing the spray speed rate at high temperatures will cause rapid deposition of a larger number of aerosol droplets, leading to produce a more agglomerated porous microstructure. Moreover, the effect of spray speed is enhanced between 560-580°C. Therefore, if the speed rate and temperature are in the optimal range with respect to each other an almost dense CVD–like layer such as the sample made at the temperature of 540°C and the speed 6 can be obtained as shown in Figure 5.

CONCLUSION

Spray pyrolysis was applied to deposit LSM cathode on YSZ substrate. Metal-organic precursors and organic solvent created a homogeneous crack-free LSM cathode film. In fact, organic route has proved to be more satisfactory than aqueous one for use in manufacturing long-life porous cathodes with.

It was also shown that the temperature and spray speed play important role in the microstructural characteristics of the deposited film. The investigations have been conducted on many samples at a wide range of temperature and spray speed. However, the most appreciable results were related to the range of temperature and spray speed discussed above. Based on microstructural analysis, increasing deposition temperature promotes formation of a more porous film microstructure. Moreover, increasing the spray speed provides less time for aerosol droplets to reach the surface leading to creation of thinner film at lower temperature and a more porous microstructure at higher temperature.

REFERENCES

[1]K. Chen, Z. Lu, N. Ai, X. Chen, J. Hu, X. Huang and W. Su, Effect of SDC-Impregnated LSM Cathodes on the Performance of Anode-Supported YSZ Films for SOFCs, *J. Power Sources,* **167,** 84-89 (2007).
[2]J. Q. Li and P. Xiao, Fabrication and Characterization of La$_{0.8}$Sr$_{0.2}$MnO$_3$/Metal Interfaces for Application in SOFCs, *J. Euro. Ceram. Soc.,* **21,** 659-668 (2001).
[3]W. Wang and S. P. Jiang, A mechanistic study on the activation process of (La, Sr)MnO$_3$ electrodes of solid oxide fuel cells, *Solid State Ionics,* **177,** 1361-1369 (2006).
[4]C.-J. Li, C.-X. Li, and M. Wang, Effect of Spray Parameters on the Electrical Conductivity of Plasma-Sprayed La$_{1-x}$Sr$_x$MnO$_3$ Coating for the Cathode of SOFCs, *Surface & Coatings Technology,* **198,** 278-282 (2005).
[5]A. N. Grundy, B. Hallstedt and L. J. Gauckler, Assessment of the La-Sr-Mn-O System, *Computer Coupling of Phase Diagrams and Thermochemistry,* **28,** 191-201 (2004).
[6]D.S. Li, G. Saheli, M. Khaleel M, H. Garmestan, Microstructure Optimization in Fuel Cell Electrodes Using Materials Design, *CMC-Computers, Materials & Continua,* **4,** 31-42 (2006).

[7]D.S. Li, G. Saheli, M. Khalee, H. Garmestani, Quantitative Prediction of Effective Conductivity in Anisotropic Heterogeneous Media Using Two–point Correlation Functions, *Computational Materials Science*, **38**, 45-50 (2006).

[8]A. Wijayasinghe, B. Bergman and C. Lagergren, LiFeO$_2$-LiCoO$_2$-NiO Materials for Molten Carbonate Fuel Cell Cathodes. Part I: Powder Synthesis and Material Characterization, *Solid State Ionics*, **177**, 165-173 (2006).

[9]Y. Liu, C. Compson and M. Liu, Nanostructured and functionally graded cathodes for intermediate temperature solid oxide fuel cells, *J. Power Sources*, **138**, 194-198 (2004).

[10]S. Zha, Y. Zhang and M. Liu, Functionally Graded Cathodes Fabricated by Sol-Gel/Slurry Coating for Honeycomb SOFCs, *Solid State Ionics*, **176**, 25-31 (2005).

[11]G.-j. Li, Z.-r. Sun, H. Zhao, C.-h. Chen and R.-m. Ren, Effect of Temperature on the Porosity, Microstructure, and Properties of Porous La$_{0.8}$Sr$_{0.2}$MnO$_3$ cathode Materials, *Ceram. Int.*, **33**, 1503-1507 (2007).

[12]S. Bebelis, N. Kotsionopoulos, A. Mai, D. Rutenbeck and F. Tietz, Electrochemical Characterization of Mixed Conducting and Composite SOFC Cathodes, *Solid State Ionics*, **177**, 1843-1848 (2006).

[13]J. Will, A. Mitterdorfer, C. Kleinlogel, D. Perednis and L. J. Gauckler, Fabrication of Thin Electrolytes for Second-Generation Solid Oxide Fuel Cells, *Solid State Ionics*, **131**, 79-96 (2000).

[14]D. Montinaro, V. M. Sglavo, M. Bertoldi, T. Zandonella, A. Arico, M. Lo Faro and V. Antonucci, Tape Casting Fabrication and Co-Sintering of Solid Oxide "Half Cells" with a Cathode-Electrolyte Porous Interface, *Solid State Ionics*, **177**, 2093-2097 (2006).

[15]D. Perednis, O. Wilhelm, S. E. Pratsinis and L. J. Gauckler, Morphology and Deposition of Thin Yttria-Stabilized Zirconia Films Using Spray Pyrolysis, *Thin Solid Films*, **474**, 84-95 (2005).

[16]A. Kumar, P. S. Devi,w A. D. Sharma, and H. S. Maiti, Effect of Metal Ion Concentration on Synthesis and Properties of La$_{0.84}$Sr$_{0.16}$MnO$_3$ Cathode Material, *J. Power Sources*, **161**, 79-86 (2006)

[17]Z. V. Marinkovic, L. Mancic, J. F. Cribier, S. Ohara, T. Fukui and O. Milosevic, Nature of Structural Changes in LSM-YSZ Nanocomposite Material During Thermal Treatments, *Materials Science and Engineering A*, **375-377**, 615-619 (2004).

[18]H. S. Kang, J. R. Sohn, Y. C. Kang, K. Y. Jung and S. B. Park, The characteristics of nano-sized Gd-doped CeO$_2$ particles prepared by spray pyrolysis, *J. Alloys and Compounds*, **398**, 240-244 (2005).

EXAMINATION OF CHROMIUM'S EFFECTS ON A LSM/YSZ SOLID OXIDE FUEL CELL CATHODE

T.A. Cruse, M. Krumpelt, B.J. Ingram
Argonne National Laboratory
9700 South Cass Avenue
Argonne IL, 60439

S. Wang, P.A. Salvador
Carnegie Mellon University
149 Roberts Eng. Hall
Pittsburgh, Pennsylvania 15213-3890

ABSTRACT

The work presented here was conducted to improve the understanding of the mechanism by which chromium affects solid oxide fuel cell cathodes. Anode (Ni/8YSZ) supported single cells (8YSZ electrolyte and LSM/8YSZ active cathode) were purchased from InDEC and operated under various conditions; 700°C and 800°C at several current densities. E-Brite® interconnects were used for current collection on the cathode, and provided the source of chromium. Scanning electron microscopy (SEM) and transmission electron microscopy (TEM) were used to examine chromium's interactions, phases and distribution. Decreasing temperature and/or increasing the applied current results in a decrease in cell operating potential and an increase in rate of cell potential degradation. SEM analysis shows different chromium distribution under the channels of the interconnect compared with under the ribs (point of physical contact). SEM and TEM analysis shows the influence of the operating conditions on the degradation of the cathode microstructure by chromium. The operating cell potential, which is influenced by several factors, affects the degradation rate of solid oxide fuel cells in the presence of volatile chromium species.

INTRODUCTION

Solid oxide fuel cells (SOFC's) are efficient devices for producing electricity from a variety of fuels. Presently, cells based on yttria-stabilized zirconia electrolytes (~8 mol% Y_2O_3) are being targeted to operate near 800°C, with the possibility of operating as low as 650°C for the "intermediate" temperature range. Operating at lower temperatures should reduce adverse interactions between components and allow for a wider selection of materials for use in stack design and construction. One advantage of lower operating temperatures (below 1000°C) has been the introduction of metallic bipolar plates/interconnects. Replacing the $(La,Sr)CrO_3$ based interconnects with alloys (primarily ferritic stainless steels) results in reductions in the costs of materials, processing and fabrication; other advantages include higher electrical and thermal conductivities. The draw back to the use of alloys is the formation of protective oxide scales that reduce the electrical conductivity. These scales can be based on different oxides such as aluminum, which has a very high electrical resistance, or chromium, which has a much lower electrical resistance. While chromium based scales have advantages over other scales, a major concern is the volatilization of chromium from the scale and the subsequent deposition within the cathode, resulting in a decline in cell performance.

There are two types of ferritic stainless steels being considered for SOFC interconnect applications: those forming a chromia protective layer (e.g., E-Brite ® and 430) and those that form an outer $(Mn,Cr)_3O_4$ spinel layer over an inner chromia layer (e.g., 446 and Crofer 22 APU). There are also a number of alloys with small amounts of manganese, such as 441, that may form a small amount of spinel on a primarily chromia scale. Silicon is another trace element that is present in almost all steels. For interconnect application its presence can cause significant problems by the formation of a high resistive silica sub-scale under the chromia, significantly increasing the electrical resistance.

The mechanism and rate of chromium volatilization from chromia, or a chromia scale, have been examined by a number of different papers. Equations 1 and 2 show the volatilization mechanisms[1].

$$Cr_2O_3 + \frac{3}{2}O_2 \longrightarrow 2CrO_{3(g)} \qquad\qquad 1$$

$$Cr_2O_3 + 2H_2O + \frac{3}{2}O_2 \longrightarrow 2CrO_2(OH)_{2(g)} \qquad\qquad 2$$

There are several factors that can affect the partial pressure of the various chromium species. Ebbinghaus provides detailed examinations of this[2]. The oxygen and water partial pressures, temperature, and chromium source affect the partial pressure of the hexavalent chromium. Under normal SOFC operating conditions for the cathode (i.e., 650-800°C and flowing air) the P_{H2O} is the most critical factor governing chromium volatilization. For example, 3% water can raise the partial pressure of hexavalent chromium species (in the form of $CrO_2(OH)_2$, see Eqn. 2) by at least an order of magnitude relative to dry air (where CrO_3 is the primary volatile hexavalent chromium species). There are a number of references that examine and compare the different factors[1-9]. The nature of the protective oxide scale, or lack there of, can also significantly influence the rate of volatilization by about an order of magnitude (Cr_2O_3 releasing more than $MnCr_2O_4$). Other factors, such as contaminates, can also play a roll; potassium can produce a very volatile K_2CrO_4 vapor[10-11].

Once the chromium vaporizes it travels through the cathode and is reduced from the hexavalent vapor to a trivalent solid, thereby blocking oxygen reduction and efficient cell operation. There has been debate in the literature as to whether this is a chemical or electrochemical reaction. An electrochemical reduction mechanism proposed by Hilpert et al. can be seen in equation 3[1].

$$2CrO_2(OH)_{2(g)} + 6e^- \longrightarrow Cr_2O_3 + 3O^{2-} + 2H_2O \qquad\qquad 3$$

At the active triple phase boundary (TPB) the chromium reacts according to Eqn. 3, forming chromia. As the chromia grows it effectively blocks oxygen reduction at the TPB thereby preventing the cathode reaction from occurring[1,12]. In contrast, Jaing et al. has proposed a chemical redox mechanism based on the natural thermal reduction of Mn in LSM from Mn^{3+} and Mn^{4+} to Mn^{2+}, which migrates onto the YSZ surface. It subsequently acts as a reduction agent for volatile hexavalent chromium forming a Mn-Cr-O nucleus that grows as either a $(Mn,Cr)_3O_4$ spinel or chromia, equation 4.

$$CrO_2(OH)_{2(g)} + Mn^{2+}_{onYSZ} \longrightarrow Mn\text{-}Cr\text{-}O_{nuclei} + CrO_2(OH)_{2(g)} \longrightarrow$$
$$Cr_2O_3 + CrO_2(OH)_{2(g)} + Mn^{2+}_{onYSZ} \longrightarrow (Mn,Cr)_3O_{4(onYSZ)} \qquad\qquad 4$$

The $(Mn,Cr)_3O_4$ is postulated to block oxygen migration from the surface of the LSM to the electrolyte, thus preventing the fuel cell from operating[13-18].

Both proposed mechanisms result in the blocking of the TPB and result in degradation of the SOFC performance. Additionally, the free energies of formation favor a reaction between LSM and deposited chromia forming $LaCrO_3$ and $MnCr_2O_4$[19-20].

This paper examines how operating conditions such as: operating potential/applied current and temperature influence the performance of these fuel cells and how these factors effect the rate of degradation of these cells by chromium. By understanding how these operational factors influence the degradation mechanism, possible solutions and optimum operating conditions can be developed.

EXPERIMENTAL PROCEDURES

Anode supported fuel cells, type ASC1, and were purchased from InDEC® (H.C. Stark). As received cells consists of a NiO-8YSZ anode (600μm), an 8YSZ electrolyte (5μm), an LSM/8YSZ active cathode (15μm) and an LSM contact cathode (15μm). The cells had a diameter to 40mm and the cathode had a diameter of 36mm.

On the cathode side LSM contact paste was used between the cathode and interconnect (e.g., gold or E-Brite) and gold paste was used between the interconnect and gold leads. The interconnects were 15.84 mm by 16.60 mm by 3.2 mm with 5 equally spaced 1.6mm by 1.6mm air flow channels. Stainless steel interconnects were polished to a 600 grit finish, heat treated in air for 1 minute at 900°C then cleaned using a dilute pickling solution of nitric and hydrofluoric acids. Electrical contact to the anode was by nickel mesh and leads with a NiO contact paste (reduced to Ni during operation). The test fixtures were produced using alumina components and were based on previous design[21]. The cell and cathode housing were cemented in place using Zircar AL-CEM (alumina refractory cement).

The "active area" for the cathode is estimated to be between 2.63 cm^2 (area under the interconnect) to 4.79 cm^2 (the area under the interconnect and air inlet and exhaust regions). Based on SEM/EDS analysis of the chromium distribution in the cathode an active area of 3.1 cm^2 was estimated. However, based on the performance levels for ASC1 cells reported by InDEC® (600mA/cm^2 at 0.8V and 800°C), an active area of 2.63 cm^2 will be used for the results presented in the paper, although there are probably differences between InDEC's test configurations and the ones used here.

Once the cell was assembled in the fixture it was placed in the furnace (which can test up to three cells at one time). The furnace was slowly heated to the desired temperature (700°C or 800°C) and held while the anode was reduced. The reduction of the anode was accomplished by purging with nitrogen, then 4%, 25% and 50% hydrogen in nitrogen for 20 minutes each. The fuel gas and air were then passed through a room temperature, 25°C, sparger, providing ~3% water. Fuel flow was set at 400 sccm and air flow at 140 sccm. Open circuit potential was measured for all cells and was typically 1.02V±0.02 for cells operated at 800°C and 1.04V±0.01 at 700°C.

The cells were controlled and monitored using a Solartron 1480 multi-channel potentiostat cell analyzer with a 1252A frequency response analyzer. Once the cells reached operating conditions and a stable open circuit voltage for 20 minutes an initial polarization curve was generated in the galvanodynamic mode, from 0 to 4 A (or achieving 0 volts). Initial electrochemical impedance spectra (EIS) were collected at 0 A applied current and at operating current. EIS measurements were collected over a frequency range of 0.1 to 50000 Hz. The cells were then operated in galvanostiatic mode, at desired current, for 24 hours. The standard operating current was 1.15A, although some cells were operated at 0, 0.575 and 2.3 A. Every 24 hours the galvanostatic measurement was automatically interrupted and electrochemical impedance spectrum (at operating current) and a polarization curve (potentiodynamic mode to a 0.3 V decrease in potential) were collected, the cell then returned to galvanostatic mode operation. This cycle was repeated 21 times for a total operating time of 504 hours. After the run final polarization curves were measured using potentiodynamic and galvanodynamic methods (previously mentioned conditions) and a final electrochemical impedance spectrum was generated.

The cells were cooled to room temperature, removed from the fixture and the cell surfaces were examined using SEM/EDS. The cell and interconnect were then mounted in epoxy. The specimen was then cut in half, along the center channel of the interconnect. Three cross-sections were then taken from one half; one near the front of the interconnect (air entering), one near the center and one near the back (air exhaust). These cross-sections were then remounted, polished and characterized using SEM/EDS. For some cells cross-sections were also prepared for the entire cell length to help characterize the chromium distribution. Some of the cells were sent to Carnegie Mellon University for additional characterization by TEM using a FIB sample preparation technique.

RESULTS AND DISCUSSIONS

Previously the effects of physical contact of an E-Brite® interconnect were examined by comparing gold, E-Brite® and E-Brite® with a gold contact barrier at an applied current of 1.15A at 700°C and 800°C[21].

Cells were operated at four different constant current conditions; standard condition (1.15A), zero (0 A), half (0.575 A) and double (2.3 A). The zero, half and double current cells were all tested simultaneously at either 800°C or 700°C. Figure 1 shows the performance for these cells over 500 hours. When simply monitored at open circuit voltage (OCV) and no applied current the cell showed no significant change over 500 hours. As the current was increased the initial cell potential decreased, the cell took longer to achieve equilibrium conditions, and the rate of degradation increased. The three cells operated at 800°C show a slight spike in performance around 480 hours due to problems with the nitrogen supply. The three cells operated at 700°C show a couple of changes due to problems with the air supply. Reducing the operating temperature of the cells from 800°C to 700°C resulted in significant decreases in cell performance. As the applied current was increased the cell potential decreased, time to reach a steady state condition increased, and the rate of degradation significantly increased.

Comparing the EIS data for these cells, figure 2, shows that there is little change in the high frequency intercept when changing from 800°C to 700°C. This indicates that the there is little change in the contribution from the interconnect. However, the size of the arcs increased as the temperature was reduced, indicating that a change in the electrochemical mechanisms of the fuel cell are being effected; slower charge transfer, higher electrical resistances and higher ionic resistance, most of which may be attributed to the cathode reactions. There is also a shift as a higher current is used at 800°C which may be attributed to an increase in segregation within the cathode (such as Sr^{22}), decomposition of the microstructure and a resulting decrease in available oxygen (to be discussed). Figure 3 shows the polarization curves and power densities at 800°C and 700°C for the different test conditions. Reducing the temperature from 800°C to 700°C resulted in the peak power density being cut by approximately half for a given test condition.

Examination of Chromium's Effects on a LSM/YSZ Solid Oxide Fuel Cell Cathode

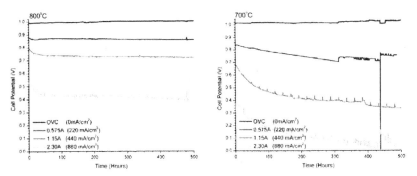

Figure 1: Effect of applied current on cell performance over 500 hours on cells operated at 800°C (left) and 700°C (right).

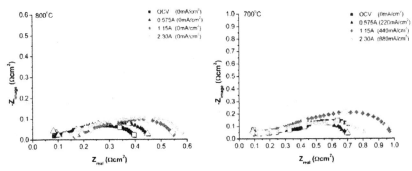

Figure 2: EIS measurements, post run all done at 1.15A applied current (as the applied current alters the measurements). Cells tested at 800C are on the left, and 700C are on the right. Open symbols correspond to a different decade of the measurement.

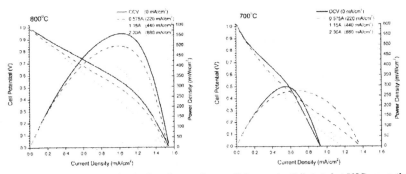

Figure 3: Polarization and power density for cells at various applied currents. Cells tested at 800C are on the left, and 700°C are on the right.

To provide an additional point for comparison, figures 4 and 5 show the performance of cells operated at 1.15A applied current after 500 hours. For both the gold interconnect and the E-Brite ® interconnect ("poisoned" cell) the cell potential begins to drop of fairly rapidly below 800°C The decrease in cell potentials in figure 4 between the different interconnects is attributed primarily to the chromium poisoning of the cathode and to a lesser extent the build of a chromia scale on the interconnect. Using the gold interconnect in figure 5 as a baseline the high frequency shift and in increase in size of the arcs shows how temperature effects the overall performance of these cells. Therefore all things being equal the only difference would be in the electrical conductivity of the scale, however the resistance of chromia only changes by a factor of 4x over this temperature range[23], which can account for some of the change, but all things are not equal there were significant changes in the microstructure of the cathode as well (to be discussed).

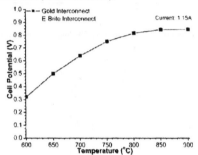

Figure 4: Cell potential taken as the temperature was reduced from 900°C to 600°C after operating for 500 hours at 800°C and 1.15A.

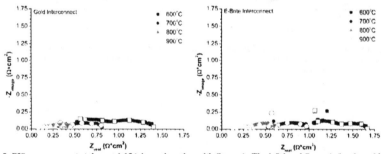

Figure 5: EIS measurements taken at 1.15A in conjunction with figure 4. The left hand figure is for the gold interconnect, the right hand figure is for the E-Brite interconnect.

Apart from the changes in cell performance and degradation rates scanning and transmission electron microscopy provided significant insight into the degradation mechanism of the LSM cathodes in the presence of chromium. Figure 6 shows a TEM image of a cell operated at 700°C, 1.15 A with a gold interconnect. This cell showed stable performance, with no degradation and operated at a cell potential of 0.73V. There is no indication of cathode decomposition, although not observed on this cell some traces of Mn on YSZ were observed with cell operated under similar conditions with an inert interconnect (E-Brite with a protective coating). However, these were only on the order of a few nanometers.

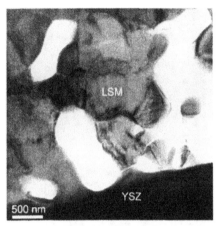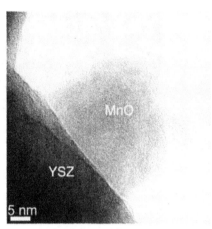

Figure 6: TEM image of cell operated at 700°C, 1.15A with gold interconnect (left) and inert interconnect (right).

When examining the microstructure of the cells operated with an E-Brite interconnect significant difference in the microstructures could be observed depending on the applied current and the operating temperature. Figure 7 shows SEM cross-sectional images of the active LSM/YSZ cathodes tested under different currents at 800°C. As the applied current was increased the amount of chromium deposited increased and the open porosity of the microstructure's porosity decreased as the LSM decomposed. The chromium contents presented in Figure 7 are averages over multiple areas of the active cathode (each area analyzed was approximately the same as the image). Comparing the areas of the active cathode directly under the interconnect ribs (in contact with the fuel cell) with those under the channels revealed significant differences in the amount of chromium deposited in the active cathode. For the cells run at 0.575 A there was approximately three to four times more chromium under the ribs than under the channels, for the cells run at 1.15 A there was approximately twice as much chromium under the ribs compared with under the channels. While at 2.3 A the chromium distribution was fairly uniform. The cells tested at 700°C had microstructures that were similar to those tested at 800°C with half the applied current (i.e. 700°C with 1.15A was similar to 800°C with 2.3A). The chromium distribution was fairly uniform in the active cathode (areas under the channels compared with areas under the ribs).

Figure 8 shows TEM images of the pores filling with an in-growth of Mn-Cr-O phase. Comparing the two images shows that less in-growth was observed for the cell operated at 800°C compared with the one operated at 700°C operated at the same applied current densities. Figure 9 shows a TEM image illustrating several types of local Cr interactions with the cathode: decomposed LSM phase, growth of a $(Mn,Cr)_3O_4$ phase from the decomposing LSM and the migration of Mn and deposition of Cr onto the surface of YSZ forming a Mn-Cr-O phase. This cell was operated at 800°C for 500 hours at 2.3 A. This microstructure is similar to that of the cell operated at 700°C for 500 hours at 1.15 A. Table I provides a comparative summary of the results of the different cell tests.

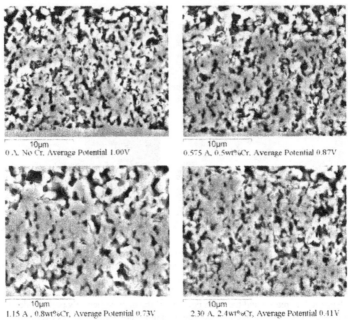

Figure 7: SEM cross-sectional images of active LSM/YSZ cathode layer operated at various currents for 500 hours at 800°C.

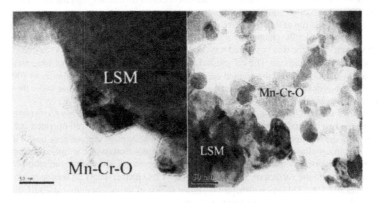

Figure 8: TEM images of LSM pores filling with Mn–Cr–O. Operated at 1.15A for 500 hours, 800°C (left) and 700°C (right).

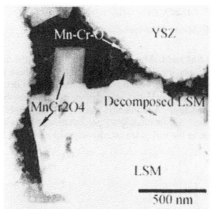

Figure 9: TEM image of decomposed LSM particle of cell operated at 800°C for 500 hours at 2.3A.

Table I: Summary of Cell Test Results

Applied Current (A)	Temperature (°C)	Cell Potential Range (V)	MnCr$_2$O$_4$ and Cr$_2$O$_3$ Observed (TEM)	MnO Observed (TEM)	LSM Decomposed (TEM)
0	700°C	1.01	---	---	---
	800°C	1.00	---	---	---
0.575	700°C	0.83-0.73	Moderate	Trace	No
	800°C	0.87-0.86	Light	No	No
1.15	700°C	0.46-0.33	High	Yes*	Yes
	800°C	0.74-0.72	Moderate	Yes	Yes
2.3	700°C	0.25-0.02			Yes (SEM)
	800°C	0.43-0.39	High	No	Yes (SEM

Light: Found on YSZ, Moderate: Found mainly on YSZ, trace on LSM, High: Found on YSZ and LSM.
*Analysis on sample without Cr source.

Figure 10 shows a relationship between the amount of chromium found in the active cathode and the average potential at which the cell operated. As the cell potential decreased (at a fixed current), resulting in a higher overpotential, the amount of chromium deposited in the active cathode increased. Based on figure 4, for the Au interconnect, there is a 22% decrease in potential going from 800°C to 700°C. For the cell with the E-Brite interconnect there is a 35% decrease in the cell potential. This difference in cell potential between the two conditions is attributed to the "poisoning" of the cathode. In addition to "poisoning" other factors also lower the cells operating potential: current density, amount of triple phase boundary (particle size, surface area, mixing of LSM and YSZ), and electrical and ionic resistances. With a decrease in the cell potential there is an increase in the driving force for the reduction of manganese from Mn^{4+}(produced by the Sr doping) and Mn^{3+} to Mn^{2+}. Using HSC to calculate the reaction to form LaMnO$_3$ (LSM is not an option) and the Nernst equation, a potential of 0.25V was determine, equation 5[19,24].

$$LaMnO_3 \longrightarrow MnO + \tfrac{1}{2} La_2O_3 + \tfrac{1}{4} O_{2(g)}$$ 5
$$\Delta G = \sim 98kJ/mol = -nFE \Rightarrow \sim 0.25V$$

One experiment examining LSM and $LaMnO_3$ showed that at 800°C the free energy of formation for LSM was ~80% of $LaMnO_3$[25]; placing the potential around 0.2V. These values correspond with the observations that cells operating at ~0.75V, and lower, showed signs of degradation in the presence of Cr, and the lower the potential the worse the degradation. While cells that were operated at 0.85V showed no indication of degradation. Additionally, Mn^{2+} is a highly mobile species and could easily migrate from the LSM onto the YSZ, facilitating the precipitation of the hexavalent chromium species; resulting in a further decrease in the cell potential, decomposition of the LSM and more precipitation of chromium.

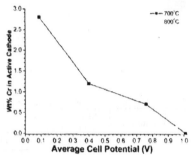

Figure 10: The average operating cell potential versus the average wt% Cr detected in the active cathode layer by EDS.

Figure 11 provides a comparison the amount of chromium found in the active cathode and the rate of degradation of the cells. This figure draws into sharp contrast how simply lowering the operating temperature by 100°C, a similar amount of chromium, can drastically increase the degradation rate of the cells. So while these LSM/YSZ cathode cells work well at 800°C with only a few percent rate of degradation in performance due to chromium poisoning, at 700°C (where the triple phase boundary area, conductivities and reaction kinetics are reduced) these cells quickly degrade to a point where they will no longer function.

Future work will focus on constant potential conditions further verify these results and determine the correlation between cell potential and degradation rates. This show help establish cell operating parameters that can minimize the effects of chromium on cell performance.

CONCLUSIONS

This work has demonstrated how cell operating conditions can influence degradation in the presence of vaporous hexavalent chromium species. It has been shown how the LSM component of the cathode can decompose, reducing the active area and closing the porous microstructure. The key driving force is the lower cell potential. The changes in temperature and applied current, both of which altered the cell potential, showed that as the cell potential decreased the decomposition of the cathode increased and the rate of degradation increased.

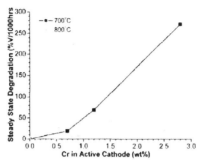

Figure 11: The average wt% Cr detected by EDS in the active cathode layer versus the steady state decline in the cells operating potential.

ACKNOWLEDGEMENTS

This work was funded by U.S. Department of Energy, Office of Fossil Energy, Solid State Energy Conversion Alliance, with Wayne Surdoval as program coordinator and Lane Wilson and Briggs White as program managers. Argonne National Laboratory is operated for the U.S. Department of Energy by UChicago Argonne, LLC. under contract DE-AC02-06CH11357.

J. Stevenson of Pacific Northwest National Laboratory for providing some of the materials tested in this experiment.

REFERENCES

[1] K. Hilpert, D. Dos,M. Miller, D. H. Peck and R. Weiß, Chromium Vapor Species over Solid Oxide Fuel Cell Interconnect Materials and Their Potential for Degradation Processes, *J. Electrochem. Soc.*, **143**, 3643-3647 (1996).

[2] B.B. Ebbinghaus, Thermodynamics of Gas Phase Chromium Species: The Chromium Oxides, the Chromium Oxyhydroxides, and Volatility Calculations in Waste Incineration Processes, *Combustion and Flame*, **93**, 119-137 (1993).

[3] C. Ginddorf, L. Singheiser and K. Hilpert, Vaporization of chromia in humid air, *J. Phys. Chem. Solids*, **66**, 384-387 (2005).

[4] E.J. Opila, D.L. Myers, N.S. Jacobson, I.M.B. Nielsen, D.F. Johnson, J.K. Olminsky and M. D. Allendorf, Theoretical and Experimental Investigation of the Thermochemistry of $CrO_2(OH)_2(g)$, *J. Phys. Chem. A*, **111**, 1971-1980 (2007).

[5] J.W. Fergus, Metallic interconnects for solid oxide fuel cells, *Mat. Sci. Eng. A*, **397**, 271-283 (2005).

[6] D.J. Young and B.A. Pint, Chromium Volatilization Rates from Cr_2O_3 Scales into Flowing Gases Containing Water Vapor, *Oxid. Metals*, **66**, 137-153 (2006).

[7] T. Brylewski, T. Maruyama, M. Nanko and K. Przybylski, TG Measurements of the Oxidation Kinetics of Fe-Cr Alloy with Regards to its Application as a Separator in SOFC, *J. Thermal Analysis Calorimetry*, **55**, 681-690 (1999).

[8] G.C. Fryburg, F. J. Kohl and C.A. Stearns, Enhanced Oxidative Vaporization of Cr_2O_3 and Chromium by Oxygen Atoms, *J. Electrochem. Soc.*, **121**, 952-959 (1974).

[9] M. Stanislowski, E. Wessel, K. Hilpert, T. Markus, and L. Singheiser, Chromium Vaporization from High-Temperature Alloys I. Chromia-Forming Steels and the Influence of Outer Oxide Layers, *J. Electrochem. Soc.*, **154**, 4, A295-A306 (2007).

[10] R.D. Brittain, K.H. Lau and D.L. Hildenbrand, Mechanism and Thermodynamics of the Vaporization of K_2CrO_4, J. Electrochem. Soc., **134**, 2900-2904 (1987).

[11] B.J. Ingram, T.A. Cruse, M. Krumpelt, Potassium-Assisted Transport in Solid Oxide Fuel Cells, J. Electrochem. Soc., **154**, B1200-B1205 (2007).

[12] S. Taniguchi, M. Kadowaki, H. Kawamura, T. Yasuo, Y. Akiyama, Y.Miyake and T. Saitoh, Degradation phenomena in the cathode of a solid oxide fuel cell with an alloy separator, J. Power Sources, **55**, 73-79 (1995).

[13] S.P. Jiang, S. Zhang and Y.D. Zhen, A fast method for the investigation of the interaction between metallic interconnect and Sr-doped $LaMnO_3$ of solid oxide fuel cells, Mat. Sci. Eng. B, **119**, 80-86 (2005).

[14] S.P. Jiang, J.P. Zhang and X.G. Zheng, A comparative investigation of chromium deposition at air electrodes of solid oxide fuel cells, J. Euro. Cer. Soc., **22**, 361-373 (2002).

[15] S.P. Jiang, J.P. Zhang, L. Apateanu and K. Foger, Deposition of Chromium Species at Sr-Doped $LaMnO_3$ Electrodes in Solid Oxide Fuel Cells I. Mechanism and Kinetics, J. Electrochem. Soc., **147**, 4013-4022 (2000).

[16] S.P. Jiang, J.P. Zhang and K. Foger, Deposition of Chromium Species at Sr-Doped $LaMnO_3$ Electrodes in Solid Oxide Fuel Cells II. Effect on O_2 Reduction Reaction, J. Electrochem. Soc., **147**, 3195-3205 (2000).

[17] S.P. Jiang, J.P. Zhang and K. Foger, Deposition of Chromium Species at Sr-Doped $LaMnO_3$ Electrodes in Solid Oxide Fuel Cells III. Effect of Air Flow, J. Electrochem. Soc., **148**, C447-C455 (2001).

[18] S.P. Jiang, Y.D. Zhen and S. Zhang, Interaction Between Fe-Cr Metallic Interconnect and $(La.Sr)MnO_3$/YSZ Composite Cathode of Solid Oxide Fuel Cells, J. Electrochem. Soc., **152**, A1511-A1517 (2006).

[19] HSC Chemistry 5.11 Software.

[20] TERRA, Edition 4.5e Phase and Chemical Equilibrium of Multicomponent Systems Software.

[21] T.A. Cruse, M. Krumpelt, B.J. Ingram, Degradation of SOFCs in Contact with E-Brite®, *Advances in Solid Oxide Fuel Cells III, Ceramic Engineering and Science Proceedings*, General Editors: Jonathan Salem and Dongming Zhu, Editior: Narottam P. Bansal **28**, 4, 289-300 (2007).

[22] B. Yildiz, K.C. Chang, D. Myers, J.D. Carter and H. You, In situ X-ray and Electrochemical Studies of the Solid Oxide Fuel Cell Electrodes, *Advances in Solid Oxide Fuel Cells III, Ceramic Engineering and Science Proceedings*, General Editors: Jonathan Salem and Dongming Zhu, Editior: Narottam P. Bansal **28**, 4, 153-164 (2007).

[23] A. Holt, P. Kofstad, Electrical conductivity and defect structure of Cr_2O_3. II. Reduced temperatures (<~1000°C), Solid State Ionics, **69**, 137-143 (1994).

[24] K.T. Jacob and M. Attaluri, Refinement of thermodynamic data for $LaMnO_3$, J. Mater. Chem., 12, 934-942 (2003).

[25] S. Tanasescu, N.D. Totir and D. I. Marchidan, Thermodynamic data of the perovskite-type $LaMnO_{3\pm x}$ and $La_{0.7}Sr_{0.3}MnO_{3\pm x}$ by a solid-state electrochemical technique, Electrochimica Acta, **43**, 1675-1681 (1998).

EVOLUTION OF NI-YSZ MICROSTRUCTURE AND ITS RELATION TO STEAM REFORMING ACTIVITY AND YSZ PHASE STABILITY

D. L. King, J.J. Strohm, P. Singh
Pacific Northwest National Laboratory

ABSTRACT

Internal steam reforming, also known as "on-anode" reforming, of methane has been extensively studied on Ni-YSZ (5Y and 8Y) cermet anode electrode of solid oxide fuel cells (SOFC). Experiments performed in "plug flow and plate" reactors indicate that the inherent activity of Ni-YSZ cermet anode remains highly dependent upon the pretreatment and reduction procedures employed. During reductive pretreatment of the anode, small Ni crystallites have been observed to exolve from the YSZ to provide a significant component to the initial enhanced catalytic activity. This activity decreases with time as the small Ni crystallites sinter. Control of this microstructure is key to controlling and stabilizing the long term reformation rates and thermal gradients along the anode. Experiments performed on Ni-5YSZ cermet anode similarly indicated initial NiO dissolution in 5YSZ and increase in the amount of undesirable monoclinic phase. The exolution of Ni crystallites during reduction results in a reduction in the monoclinic phase. With 8YSZ, nickel also dissolves and exolves from the YSZ particles but no monoclinic phase is observed. Thermal gradient management is dependent upon controlling the reforming activity along the length of the anode.

INTRODUCTION

The capability of Ni to steam reform methane and natural gas provides a unique capability of the Ni-YSZ solid oxide fuel cell anode to directly convert such fuels to electric power. Balancing the endothermic steam reforming reaction with the exothermic electrochemical oxidation of H_2 and CO provides a method to provide high thermal efficiency, reduction in excess cathode air flow, and reduction in size or elimination of a separate reformer.[1] One of the challenges is matching the reforming activity with the electrochemical activity in order to avoid unacceptably large thermal gradients along the flow axis, particularly near the fuel inlet. Such gradients, if sufficiently large, can result in cell damage.[2,3]

This work was initiated in order to quantify reforming activity of Ni-5YSZ bulk anode material currently being utilized in the SECA Core Technology Program at the Pacific Northwest National Laboratory. In the course of the reforming study, carried out under accelerated aging conditions, it was found that the activity of the reduced anode powders toward methane steam reforming showed a high initial activity, but activity declined with time until a final steady state activity was reached[4]. The initial activity as well as the activity profile with time was found to be quite dependent upon the method of pretreatment (reduction) of the sample as well as the temperature, steam content, and hydrogen content in the feed. This performance can be traced to the evolution of Ni microstructure during the course of the tests[4]. The initial Ni reforming activity derives from a combination of bulk Ni and Ni that exolves from the YSZ during the reduction. The presence of NiO in the YSZ after the high temperature calcination (sintering) can be demonstrated by several techniques, one of the most straightforward being X-ray diffraction (XRD). In the course of studying the XRD traces in more detail, it was found that NiO dissolved in the YSZ, and its exolution under reducing conditions, correlated with changes in the content of the monoclinic phase of ZrO_2. This paper describes observations on the development of Ni microstructure within the Ni-YSZ anode, and compares the effect of yttria content in the YSZ (5Y vs. 8Y) on phase stability and reforming performance.

EXPERIMENTAL

The bulk Ni-5YSZ and Ni-8YSZ anode materials were prepared from NiO (Baker) and 5YSZ or 8YSZ (Unitek) powders. The powders were first co-milled in an attrition mill using isopropanol as a solvent. Typically the resulting powder was tape cast to form a wafer, using isopropanol as solvent and carbon as a pore former, with subsequent firing in air using a temperature program of 0.5°C/min to 450°C and held for 1h followed by a 2°C/min ramp to 1375°C with a hold for 1h. We also tested powders taken after the milling process and calcined using the same protocols without the tape casting and forming steps. We found virtually no difference in the performance.

The Ni-YSZ sintered anode was ground with a mortar and pestle and a particle size between 100 and 200 mesh was used in the reactor tests. We typically employed 20-50mg of sample diluted with YSZ (1:10 anode:YSZ weight ratio). The diluted anode was then loaded into an INCONEL® (inconel) 625 reactor tube (9 mm inner diameter), and heated using a tube furnace. The temperature of the catalyst bed was monitored by directly placing a thermocouple within the catalyst bed. Pretreatment of the anode employed either a ramped reduction in hydrogen from ambient to the final reaction temperature (with 1 hour hold), or an isothermal reduction at the reaction temperature (generally 750°C) subsequent to calcination in air followed by N_2 purge and stepped introduction of hydrogen over the course of 1.5 hours. Reactant feed to the system comprised CH_4 (Matheson, UHP), H_2 (Matheson, UHP), and H_2O, and in some cases diluted with He or N_2 for operational flexibility. H_2O was vaporized and mixed with the gas feeds upstream of the reactor. Feed composition typically comprised a ratio of $H_2O/CH_4/H_2/N_2$ of 3/1/1/1. A typical feed rate gave a WHSV (CH_4) of 49g/g_{cat}-h (GHSV 324K cc/g_{cat}-h) although other conditions were sometimes used. Reaction products were analyzed online using a Hewlett Packard microGC, with dual TCD detector and column configuration analyzing H_2, CO, CH_4, and N_2/He on a molecular sieve A column (Ar carrier gas) and a PlotQ column analyzing CO_2 (He carrier gas).

A plate reactor was also used to directly measure thermal gradients and time dependent changes during reforming over a SOFC anode. For plate reactor tests, a 4 cm x 8 cm sintered Ni-YSZ tape cast planar anode 0.5 mm thick was prepared and inserted within the reactor (inconel 625). The anode plate was held 1mm from each of the walls of the inconel reactor walls using Al_2O_3 spacers allowing feed gas to flow over both sides of the wafer. The reactor was equipped with 12 thermocouples allowing measurement of thermal profiles both along the axis of flow and across the wafer. Little difference in temperature was found across the wafer. The entire assembly was supported inside a furnace in order to provide constant power to the reactor over the course of the experiment.

X-ray (XRD) characterization was carried out using a Philips Wide-Range Vertical Goniometer with an incident beam 2θ-compensating slit and soller slit, a fixed receiving slit, a diffracted beam graphite monochromator, and a scintillation counter detector. The X-ray source (Philips XRG3100) operated a fixed-anode, long-fine-focus Cu tube at 1800W. A wide-range scan was performed on all samples using a 2θ step of 0.05 from 20 to 85° with 1 sec steps. For quantification of the monoclinic phase and ratio of cubic and tetragonal, a high resolution scan from 20 to 36° and 72 to 77°, respectively, was performed using a 2θ step of 0.02 and 5 sec dwell. The fraction of monoclinic phase was determined using equation 1, and equation 2 was used to determine the cubic and tetragonal phases[5]. All samples were ground by a mortar and pestle to <350 mesh and loaded on a quartz sample holder. Prior to each set of analysis, silicon (NIST SRM 640b) was used as an external d-spacing standard.

$$X_m(\%) = \frac{I(111)_m + I(11\overline{1})_m}{I(11\overline{1})_m + I(111)_m + I(111)_{c,t}} \times 100 \qquad (1)$$

$$X_c(\%) = (100 - X_m) \frac{I(400)_c}{I(400)_c + I(004)_t + I(400)_t} \qquad (2)$$

Samples for TEM analysis were prepared by pressing finely ground powders onto a Cu and carbon grid. HRTEM analysis was carried out on a Jeol JEM 2010F microscope with a specified point to point resolution of 0.194nm. An operating voltage of 200 keV was used and all images were digitally recorded with a slow scan charge coupled device camera (image size 1024 x 1024 pixels). Image processing was carried out using a Digital Micrograph (Gatan). Qualitative atomic compositions were determined by EDS analysis performed on a Vacuum Generators HB603 dedicated scanning transmission electron microscope operating at 300 keV with a 30 μm virtual objective aperture. Characteristic X-rays were analyzed using an Oxford Instruments windowless Si(Li) EDS detector with an incident probe size of less than 1 nm and beam current of 0.5nA.

RESULTS
Powder Activity Tests
Figure 1 shows a typical reformation activity vs. time curve for Ni-5YSZ operating at 700°C and a S/C/H₂ ratio of 3/1/1 that was pre-reduce via a ramped reduction. The space velocity is high, ensuring less than full conversion of the methane and facilitating observation of deactivation effects. After approximately 120 hours on stream, the catalyst activity stabilized at a value well below the initial activity. Figure 2 shows the morphology of a YSZ particle surface following initial ramped reduction (held at 700°C for 1h) and subsequent to steam reforming. Small particles of Ni metal can be seen, the result of their exolution from YSZ following the reduction procedure. Following reforming at 700°C for 145h, these particles have sintered, which helps to explain the loss of activity with time demonstrated in Figure 1. This is not the whole story, however, as shown in Figure 3. In a separate experiment, the pure bulk NiO source material was milled and sintered in analogous fashion to the Ni-YSZ, but without any YSZ present, and its reforming activity with time is shown and compared with Ni-YSZ. In this case the comparison was carried out at 750°C, which may help to account for the decreased time to reach lined out activity. It is clear that both bulk Ni and small crystal Ni contribute to the methane reforming activity, at comparable activity levels, and both exhibit deactivation with time as a result of sintering.

The activity of the Ni-YSZ and deactivation profile is somewhat different when a different pretreatment and reduction protocol is used. In order to simulate initial firing of an assembled cell, the Ni-YSZ was heated in air at 800°C (to simulate glass sealing) and subsequently cooled to 750°C in nitrogen and reduced at 750°C isothermally using a staged introduction of H₂. Comparison between the two different reduction protocols on the reforming activities is provided in Figure 4. It appears that the activity of the Ni-YSZ anode material is quite similar to the bulk Ni (shown in Figure 3). In this case there does not appear to be any activity provided by the exolution of Ni from the NiO dissolved in the YSZ. However, there is a specific aspect of the activity trace at ~15 hours on stream that differs from the bulk Ni: a small discontinuity occurs, and activity rises slightly before leveling out. This phenomenon is reproducible. The TEM analysis of this sample is provided in Figure 5. The initially reduced sample shows no evidence of small Ni particles being present at the surface of the YSZ particle, in contradistinction with the samples prepared via ramped reduction. The isothermally reduced sample, following stream reforming, shows rather unusual surface features, which include spots or "pock" marks of <5nm in diameter. In some cases these spots appear crater-like. EDS analysis indicates that the surface of this material following the reaction is significantly Ni enriched, although the oxidation state of the nickel cannot be established. It is possible that the slight increase in activity observed after 15 hrs, for the isothermally reduced NiYSZ, is due to the slow surface enrichment of Ni exolving from the YSZ during methane reforming.

In this work, 5YSZ was employed as the bulk anode material as it is representative of materials being employed in other fuel cell studies at PNNL. However, since much work by others describes the use of 8YSZ for the anode, the differences in performance between Ni-5YSZ and Ni-8YSZ were also examined. For the ramped reduction procedure, show in Figure 6, initial and final activities are similar, although the deactivation profile is slower with 8YSZ, perhaps indicating some small difference in interaction between NiO and the 8YSZ. With the isothermally reduced samples, Figure 7, there is slightly greater overall activity with the 8YSZ throughout the run, but in general the performance is quite similar, indicating that the yttria content is not an important factor in determining the overall reforming activity.

Plate Reactor Tests

A plate reactor was assembled which allowed the use of a Ni-YSZ wafer approximately 40x80x0.5mm for evaluation in methane steam reforming. The reactor was equipped with thermocouples at various locations along the flow axis (in the direction of fuel flow) in order to measure the extent and location of thermal gradients. Feed gas flowed across both sides of the anode wafer in 1mm flow channels. The large mass of the inconel reactor and its close distance to the wafer led to its functioning as a heat sink. This precluded obtaining true representative measurements of temperature as a result of reforming. The trend, however, is believed to accurately monitor the location and change of the thermal profiles. Figure 8 shows the thermal profile versus time for a material initially activated via an isothermal reduction at 750°C. During the test, constant external heating was provided to the unit (the entire unit was placed inside a heated furnace). Initial temperature was set using the same gas flow rates but with inert gas feed. The reaction was carried out at 750°C and a $S/C/H_2/N_2$ ratio of 3/1/4/1, simulating a partial anode reforming situation with some recycle of anode tail gas (but with N_2 representing CO). The position of the largest endotherm migrates down the flow axis with time. This is the result of a decrease in catalytic activity of the Ni-YSZ, first at the front edge of the wafer and subsequently down the flow axis. The magnitude of this endotherm is dampened with time, indicating that the locus of activity is actually spread over the length of the anode rather than being focused at a specific location. At all times during the test methane conversion remained near equilibrium. At the end of the experiment, the wafer was removed and activity again measured using the powder test procedures described earlier. As shown in Table I, despite the long period of operation in the plate reactor (865 hrs), additional deactivation was observed with time during the powder test. Catalyst samples taken from both the leading and back edges of the plate showed similar activity profiles, and by comparison it can be seen that the activity is substantially lower than that obtained with a fresh Ni-YSZ sample. Clearly, a substantial amount of catalyst deactivation has taken place during the plate reactor test, such that the activity is slightly lower than the Ni-YSZ catalyst exposed to powder testing only.

XRD and Phase Stability Studies With Ni-5YSZ

X-ray diffraction (XRD) methods were also employed to help verify the dissolution of NiO in the YSZ following calcination, and to confirm that this dissolved NiO is the source of the small Ni crystallites that appear at the surface of YSZ following reduction. This is shown in Figure 9. The position of the $(111)_{c,t}$ ZrO_2 reflection is monitored following several different pretreatments. Compared with pure YSZ, an increase in the reflection angle following calcination indicates a decrease in d-spacing, consistent with the addition of NiO reducing the lattice parameter/spacing. Ramped reduction at 700°C in H_2 results in a movement of the $(111)_{c,t}$ peak to lower 2θ, consistent with some loss of NiO (a consequence of reduced Ni crystallites appearing at the surface). Prolonged reduction at 750°C appears to exolve even more Ni out of the YSZ. On the other hand, 800°C reduction in the presence of a steam-rich environment indicates that a greater fraction of NiO remains in the YSZ.

Thus, even under a net reducing environment, it evident that the solubility of NiO in the YSZ depends on the H_2O partial pressure and perhaps other conditions.

During the course of studying the XRD traces for the movement of the $(111)_{c,t}$ ZrO_2 reflection, a small change in the XRD spectrum indicated the presence of a monoclinic phase. This phenomenon was further investigated. Table II compares the phase composition of the ZrO_2 following various pretreatments. The 5YSZ, as-received or milled with NiO, shows the presence of approximately 23% monoclinic zirconia, and the balance is the cubic phase. Following sintering at 1375°C of the as received 5YSZ, virtually all the monoclinic phase disappears, with the formation of a combination of cubic and tetragonal phases. However, when the NiO is present, following the calcination some monoclinic phase remains, approximately 2-4% compared with <1% with 5YSZ without NiO.

After reduction using the ramped reduction protocol, virtually all the monoclinic phase is removed, with a combination of tetragonal and cubic phases remaining. On the other hand, with the isothermal reduction procedure the amount of monoclinic phase remains essentially the same as with the original calcined sample. This implies that during the exolution of Ni from the YSZ lattice the monoclinic phase is destabilized and transforms to the cubic and tetragonal phases.

Samples that had undergone extended testing in the plate reactor at 700°C were also evaluated. The full wafer had been pre-treated using a ramped hydrogen reduction, hence initially there should have been very low monoclinic phase present. Samples from both the front and back edges of the wafer show that significant levels of a monoclinic phase have formed—approximately 5% at the leading edge and 10% at the back edge. This 10% can be lowered to approximately 5% following subsequent powder testing. Similar but slightly different results were obtained with samples obtained from plate reactor tests at 750°C using the isothermal reduction method. In this case, the leading edge shows comparable levels of monoclinic phase as the starting material, suggesting little migration of Ni-NiO in or out of the 5YSZ. On the other hand, there is a significant increase of monoclinic phase at the back edge sample, >11%.

Phase Stability of 8YSZ

Due to the appearance of the monoclinic phase with Ni-5YSZ, the effect of NiO dissolution into and exolution from 8YSZ was also examined. The reactor tests previously discussed in Figure 5 indicated catalytic performance with time similar to 5YSZ, indicating that dissolution and exolution of NiO-Ni also occurs in Ni-8YSZ. Table II shows the XRD analysis for Ni-8YSZ, following calcination, both isothermal and ramped reduction, and reaction. It is clear that the 8YSZ behaves much differently from 5YSZ, in that in no case is there evidence for the appearance of the monoclinic or tetragonal phases. The higher concentration of yttria appears to stabilize the ZrO_2 cubic phase under the conditions examined.

DISCUSSION

This work shows that the performance of Ni-YSZ is dependent on the microstructure that develops as a result of pretreatment protocols and conditions of the reforming operation. There are two components to the activity of Ni-YSZ--that provided by the bulk material and that provided by Ni that exolves from YSZ during reduction and reaction. Somewhat surprisingly, ramped reduction generates a different microstructure from isothermal reduction. The ramped reduction protocol generates small Ni crystallites at the surface of the YSZ, whereas no such crystallites are observed following the isothermal reduction. However, following reaction with the material that has undergone an isothermal reduction, there is clearly some Ni surface enrichment, indicating that the dissolved NiO may still participate in reaction. A primary phenomenon that takes place in all reforming tests involves sintering of the Ni, and it appears to affect both types of Ni particles. However, it also appears that some NiO remains within the YSZ or near the YSZ surface, even under net reducing conditions. This dissolved NiO can provide a reservoir of available Ni that may be tapped, or under steady state it may simply remain dissolved in the YSZ. As an example, front and back end powder samples obtained from long term steady state plate reactor tests, following a new reductive pretreatment, show a slight increase in activity. This indicates an increase in Ni availability that most likely came from the NiO in the YSZ. However, this increase is always followed by a loss in activity until lineout is achieved. This is probably the result of subsequent Ni sintering, although re-dissolution of Ni into the YSZ as the oxide may also be occurring to lower the activity. Although outside the scope of this paper, other studies performed indicate that the Ni microstructure is also dependent, not only on the reduction protocol and aging, but also the local atmosphere and reaction conditions. This indicates that the Ni (bulk and Ni/NiO on or within the YSZ) will reach a stable microstructure that is dependent upon local gas composition, temperature, reduction protocol, and exposure time.

The phase stability studies for Ni-5YSZ show a parallel performance with the reactor activity tests. For example, the Ni-5YSZ sample obtained by ramped reduction resulted in the appearance of Ni at the surface of the YSZ, higher initial activity, and reduction of NiO content within the YSZ. This appearance of Ni on the YSZ surface, and therefore disappearance of NiO from the YSZ, correlated with a decrease in the monoclinic phase. On the other hand, the sample prepared by isothermal reduction showed no small Ni crystallites, lower initial activity, and less apparent loss of NiO from the 5YSZ. This correlated with retention of the monoclinic phase within the YSZ. These results indicate that the presence of dissolved NiO correlates with the presence of the monoclinic phase of 5YSZ, and its removal via a reduction process leads to loss of monoclinic phase.

Although exolution of Ni from 5YSZ correlates with a decrease in the monoclinic phase, it is less clear whether an increase in the monoclinic phase as a result of reaction correlates quantitatively with NiO re-dissolution into YSZ. For example, Ni-YSZ following calcination and subsequent isothermal reduction showed approximately 3.8% monoclinic phase, and reactor tests suggested that no Ni had exolved from the YSZ during the reduction pretreatment. In plate reactor tests using isothermally reduced Ni-5YSZ, post analysis showed that the back end wafer samples had monoclinic phase of about 11.3%. It is difficult to understand how so much additional NiO became available for dissolution, unless it came from the bulk NiO. Alternatively, it is possible that NiO facilitates the formation of the monoclinic phase, but production of the monoclinic phase may be catalytic rather than stoichiometric. Regardless, it is clear that the monoclinic phase is present and can increase during the internal reforming process.

It remains to be determined whether the appearance of the monoclinic phase has an adverse affect on cell performance. It has been established previously that the presence of monoclinic YSZ can affect the mechanical and electrical properties of YSZ at temperatures below $400^{\circ}C^6$, but there is little available literature on the effect of the monoclinic phase on fuel cell bulk anode performance or stability. Long term cell testing under internal reforming conditions is indicated. Certainly a safe choice of operation would be to utilize 8YSZ in place of 5YSZ, which showed maintenance of the cubic phase under all test conditions examined in this study. Indeed, the presence of dissolved NiO has been shown to stabilize the cubic relative to the tetragonal phase at $1000^{\circ}C^7$.

Finally, some comment should be made regarding the anode activity and the generation of unacceptable thermal gradients. In separate work (not shown here), it was also established that under less accelerated aging conditions deactivation also occurred, although at a much slower rate. Thus, for some initial period, gradients may be large and unacceptable. Long term sintering and dissolution of Ni will result in a substantially less active anode, with activity declining by an order of magnitude or more. Although it is difficult to predict performance without active anode testing, rough calculations suggest that with a steady state anode such as generated in our studies, the thermal gradients will be much smaller and likely manageable. Therefore, intentional pre-sintering could be part of an initial protocol to bring the anode activity to a steady state where thermal gradients are minimized while maintaining complete methane reforming over the length of the cell.

CONCLUSION

The activity of Ni-YSZ anodes toward methane internal reforming derives from two sources of Ni: bulk Ni derived from the NiO source and from smaller Ni particles that may exolve from YSZ during reduction. Significant quantities of small Ni crystallites appear following a ramped reduction, but are not evident when an isothermal reduction is employed. Under steam reforming conditions, both types of Ni particles undergo sintering and both types contribute to overall activity. Although sintering was rapid in our experiments, which employed high space velocities, sintering (and therefore anode catalytic deactivation) would be much slower under conditions actually employed during fuel cell operation. Thus, initial anode activity could be unacceptably high, although with time the sintering would result in an anode that would likely show only modest thermal gradients along the flow axis. A comparison of Ni-5YSZ and Ni-8YSZ showed that there was little difference in catalytic performance, either initially or following aging. However, the Ni-8YSZ sample showed only the cubic phase either before or after reaction, whereas the Ni-5YSZ sample showed that some monoclinic phase was generated under both conditions. Since Ni-5YSZ is typically employed only in the bulk anode rather than the active layer, the presence of some monoclinic phase may not be problematic but deserves further investigation.

REFERENCES

[1] K. Ahmed and K. Foger, Kinetics of Internal Steam Reforming of Methane on Ni/YSZ-based Anodes for Solid Oxide Fuel Cells, *Catalysis Today,* **63**, 479-487 (2000).

[2] J. Meusinger, E. Reinsche, and U. Stimming, Reforming of Natural Gas in Solid Oxide Fuel Cell Systems, *J. Power Sources.* **71**, 315-320 (1998).

[3] M. Boder and R. Dittmeyer, Catalytic Modification of Conventional SOFC Anodes with a View to Reducing Their Activity for Direct Internal Reforming of Natural Gas, *J. Power Sources* **155**, 13-22 (2006).

[4] J. J. Strohm, D. L. King, X. Wang, P. Singh, K. Recknagle, and Y. Wang, Effects of Activity and Deactivation of Ni-YSZ SOFC Anodes on Thermal Profiles During Internal Reforming of Methane, *Prepr. Pap.- Am. Chem. Soc., Div. Fuel Chem.* **52**(2), 607-609 (2007)

[5] R. C. Garvie and P. S. Nicholson. Phase Analysis In Zirconia Systems, *J. Am. Ceram. Soc.,* **55**(6), 303-305 (1972).

[6] X. Guo. Property Degradation Of Tetragonal Zirconia Induced By Low-Temperature Defect Reaction With Water Molecules. *Chem. Mater.,* **16**(21), 3988-3994 (2004).

[7] H. Kondo, T. Sekino, T. Kusunose, T. Nakayama, Y. Yamamoto, and K. Niihara, Phase Stability and Electrical Property of NiO-Doped Yttria-Stabilized Zirconia, *Mater. Lett.* **57**, 1624-1628 (2003).

Table I. Summary of the Activity of Ni-YSZ Materials After Various Pretreatments and Reforming Conditions

Material	Reduction Procedure	TOS (hr)	Conversion (%)	rate (mol_{CH_4}/g_{cat}-hr)	rate (mol_{CH_4}/g_{Ni}-hr)	rate (mol_{CH_4}/mol_{Ni}-hr)
Ni-5YSZ	Ramped	0 hr	75.5	3.67	7.26	426
		43 hr	11.9	0.32	0.64	38
Ni-8YSZ	Ramped	0 hr	69.8	3.13	6.20	364
		44 hr	11.6	0.31	0.61	36
Bulk Ni*	Ramped	0 hr	44.6	2.20	2.20	129
		20 hr	7.0	0.33	0.33	19
Ni-5YSZ	Isothermal	0 hr	48.2	1.87	3.53	224
		18 hr	7.3	0.20	0.38	22
Ni-8YSZ	Isothermal	0 hr	59.8	2.51	4.98	292
		20 hr	10.6	0.31	0.60	35
Ni-5YSZ Plate ** Leading Edge	Isothermal	0	9.0	0.26	0.51	30
		14	5.0	0.14	0.28	16
Ni-5YSZ Plate** Back Edge	Isothermal	0	8.9	0.26	0.51	30
		14	5.3	0.15	0.30	18

* GHSV = 663K hr⁻¹, all other tests performed at GHSV=~332K hr⁻¹

** Activity Ni-5YSZ wafer after testing in plate reactor. TOS in plate reactor 865hrs

All rates are based on integral rate calculations to adjust for approach to equilibrium

Table II. Phase Composition of Ni-YSZ Materials After Various Treatments, Reduction Protocols, and Reforming Conditions

Material	Reduction Conditions		Reforming Conditions				Phase Analysis (XRD)			
	Procedure	Temp (°C)	TOS (hr)	S/C/H2/N2	Temp (°C)	WHSV (g_{CH_4}/g_{cat}-hr)	Monoclinic (%)	Cubic (%)	Tetragonal (%)	c/t
5YSZ Unitec: as received	none	none	none	none	none	none	23.1	76.9	0.0	
NiO+5YSZ milled powders	none	none	none	none	none	none	23.0	77.0	0.0	
5YSZ Unitec: calcined 1375°C	none	none	none	none	none	none	0.8	51.3	47.9	1.1
NiO+5YSZ powders: calcined 1375°C	none	none	none	none	none	none	2.4	59.2	38.3	1.5
Ni-5YSZ Anode: calcined 1375°C	none	none	none	none	none	none	3.8	58.6	37.6	1.6
Ni-5YSZ Anode: calcined 1375°C	ramped	750	none	none	none	none	0.5	53.2	46.3	1.1
Ni-5YSZ Anode: calcined 1375°C	ramped	750	43	3/1/1/1	750	49	1.4	51.3	47.3	1.1
Ni-5YSZ Anode: calcined 1375°C	isothermal	750	none	none	none	none	3.6	54.3	42.2	1.3
Ni-5YSZ Anode: calcined 1375°C	isothermal	750	18	3/1/1/1	750	49	2.2	47.0	50.8	0.9
Ni-5YSZ Plate (leading edge) 700°C test	ramped	700	135	3/1/2/0.5	700	0.8	5.0	62.9	32.1	2.0
Ni-5YSZ Plate (back edge) 700°C test	ramped	700	135	3/1/2/0.5	700	0.8	10.5	61.2	28.3	2.2
Ni-5YSZ Plate (leading edge) 750°C test	isothermal	750	865	3/1/4/1	750	1.12	3.2	58.2	38.6	1.5
after activity test	ramped	750	14	3/1/1/1	750	49	3.4	57.3	39.3	1.5
Ni-5YSZ Plate (back edge) 750°C test	isothermal	750	865	3/1/4/1	750	1.12	11.3	52.2	36.5	1.4
after activity test	ramped	750	14	3/1/1/1	750	49	5.6	47.5	46.9	1.0
8YSZ: calcined 1375°C	none	none	none	none	none	none	0.0	100.0	0.0	
Ni-8YSZ Anode: calcined 1375°C	none	none	none	none	none	none	0.0	100.0	0.0	
Ni-8YSZ Anode: calcined 1375°C	ramped	750	none	none	none	none	0.0	100.0	0.0	
Ni-8YSZ Anode: calcined 1375°C	isothermal	750	none	none	none	none	0.0	100.0	0.0	

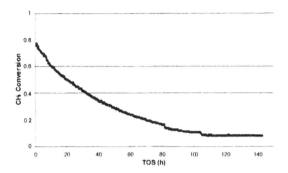

Figure 1. Activity profile of Ni-5YSZ under methane steam reforming following ramped reduction in hydrogen. Conditions: T=700°C, S/C/H = 3/1/1, WHSV = 48g_{CH4}/g_{cat}-h

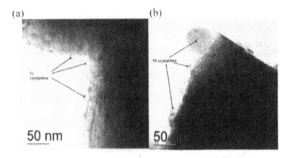

Figure 2. TEM image of Ni-YSZ following (a) ramped reduction in hydrogen at 700°C and (b) following steam methane reforming at 700°C for 65h at S/C/H = 3/1/1. Large particles are YSZ, smaller particles (5-50 nm) are pure Ni

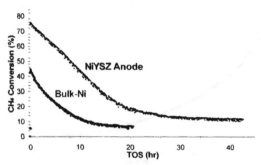

Figure 3. Comparison of methane steam reforming activity between bulk Ni and Ni-YSZ using same source and quantity of NiO. Test conditions: 750°C, S/C/H/N = 3/1/1/1, WHSV of 49 and 78.3 g_{CH4}/g_{cat}-h for the anode and bulk Ni, respectively (78.3 g_{CH4}/g_{Ni}-h for both materials).

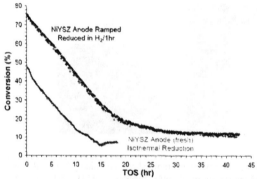

Figure 4. Effect of the reduction protocol on the methane steam reforming activity over Ni-5YSZ anode materials. Test conditions: 750°C, S/C/H$_2$/N$_2$ = 3/1/1/1, WHSV = 78.3 g_{CH4}/g_{cat}-h)

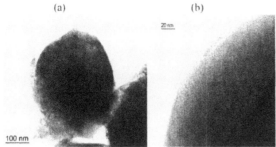

Figure 5. TEM images of Ni-YSZ following (a) isothermal reduction in hydrogen at 750°C and (b) following steam methane reforming at 750°C for 20h at S/C/H = 3/1/1.

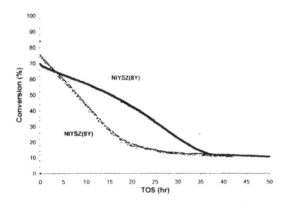

Figure 6. Comparison of 5Y and 8Y Ni-YSZ activity following ramped reduction. Conditions: T = 750°C, S/C/H = 3/1/1, WHSV = 49 g CH$_4$/g cat-h

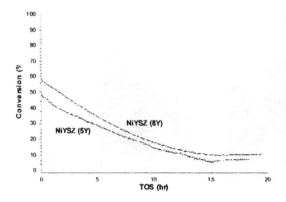

Figure 7. Comparison of 5Y and 8Y Ni-YSZ activity following isothermal reduction.
Conditions: T = 750°C. S/C/H = 3/1/1. WHSV = 49 g CH₄/g cat-h

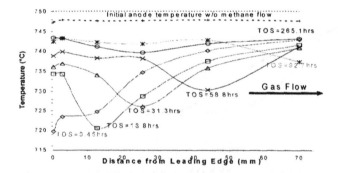

Figure 8. Thermal profiles as a function of time on stream and location for Ni-YSZ in
plate reformer test. Conditions: isothermal reduction at 750°C, nominal reaction
temperature 750°C. S/C/H = 3/1/4. GHSV = 1.1 g CH₄/g cat-h

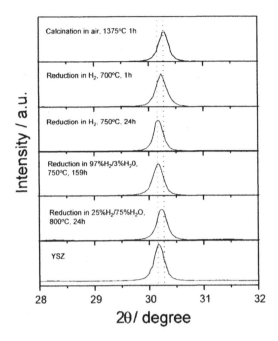

Figure 9. XRD traces of Ni-5YSZ following different treatments. All reductions were carried out using the temperature ramp procedure. The peak shows the position of the $(111)_{c,t}$ ZrO$_2$ reflection.

SYNTHESIS AND CHARACTERIZATION OF Ni IMPREGNATED POROUS YSZ ANODES FOR SOFCs

C. Anand Singh and Venkatesan V. Krishnan*
Department of Chemical Engineering, Indian Institute of Technology Delhi
New Delhi, India.
anandzing@yahoo.co.in, vvkrish@chemical.iitd.ac.in

ABSTRACT

Wet impregnation of Ni onto a preformed porous YSZ substrate has been used to fabricate the SOFC anode instead of the conventional NiO-YSZ cermet (prepared by co-sintering), with the goal of achieving higher electrical conductivity at lower Ni loadings. Two different pore formers, graphite and polystyrene were used, thus obtaining two different pore morphologies in terms of pore shape and size distribution. Cells with anodes made from graphite pore formers (Ni loadings of 9.5 vol.%) showed maximum power densities of only 50 mW/cm^2 at 850°C which is much lower than that of conventional Ni-YSZ systems. However the cell performances improved drastically once the Ni was impregnated onto polystyrene based anodes (250mW/cm^2 maximum power at 850°C for Ni loading of about 10.6 vol.%). Despite the overall conductivity of the anodes being over 50 S/cm in both cases, verified by conductivity measurements under similar conditions, it was found that the cell performance deteriorated with time for cells made with both morphologies. However the manner and extent of deterioration of performance throws light on the influence of morphology on the electrochemical reactions at the three phase boundary, as well as on the overall conductivity of the anode. Having kept the cathode constant (LSM-YSZ) in all cases, changes in performance are attributed mostly to the anode.

INTRODUCTION

The majority of anodes for Solid Oxide Fuel Cells (SOFC) are composed of a Ni and Yttria Stabilized Zirconia (Ni-YSZ cermet) and fabricated by mixing NiO and YSZ powders, followed by high temperature co-sintering (~1500°C) to generate a matrix of NiO and YSZ phases. The NiO is then reduced to produce Ni metal and in the process, porosity is generated as well. Electrical conductivity of the anode is a function of nickel content in the matrix and follows the S-shaped curve[1]. A minimum loading of 30 vol.% Ni is necessary to ensure sufficient electrical conductivity in the anode following this percolation behavior[1]. During fuel cell operation the fuel cell is exposed to reducing condition and the cermet structure is maintained. At a typical operating temperature of 800°C, nickel can reoxidize to NiO at an oxygen partial pressure, pO_2, greater than about 10^{-14} atm. For lowering reoxidation kinetics to a minimal level, temperatures of less than 500°C may be necessary, which are not feasible in an SOFC.[2] Thus, reoxidation can occur in situations like seal leakage, fuel supply interruption or system shutdown, resulting in an increase in volume due to the Ni $\rightarrow$ NiO transition. Hence thermal stresses tend to develop due to volume changes and differences in CTE's between the Ni and YSZ phases. This redox cycling occurs in commercial SOFC operations and results in large changes in bulk volume of individual phases, which may have a significant effect on the interfaces within a fuel cell and thus results in mechanical instability and consequently, performance degradation[3].

One way of improving the redox tolerance is by modification of anode microstructure. Sufficient electrical conductivity at lower Ni loadings, should lead to better redox stability and lower the chances of cell breakage due to incompatibility in CTE's (Ni-YSZ vs NiO-YSZ) caused by Ni-NiO transformation. Clemmer et al. have prepared porous Ni/YSZ composites by adding Ni coated graphite to tape casting suspension. When the graphite burns away it leaves a percolated Ni network at a much lower volume fraction than would be required from a dense Ni particle[4]. Pratihar et al. have used electroless coating of Ni on YSZ to reduce the Nickel loading[5].

Another technique for reducing the Ni loading is by incorporation of Ni into a preformed porous anode structure. Incorporation or infiltration of catalytic and conducting species into preformed porous anode structures have already been carried out by Gorte and coworkers e.g., Cu and CeO$_2$ on YSZ[6]. Infiltration, in general may be achieved by solution impregnation and calcination or via melting of salt precursors into the pores[7].

In this work, Ni-impregnation on porous anodes and SOFC's with such anodes have been investigated for their performance. The morphology of the substrates has also been varied and its role in performance characteristics, explored. In an earlier work, by the authors[8], electrical conductivity trends were investigated for porous structures, prepared using graphite pore former containing impregnated Ni. The initial data on conductivity as function of Ni vol.% was very encouraging; the authors reported an electrical conductivity of about 150 S/cm at just 10 vol.% loading on a porous YSZ anode substrate[8].

The primary difference in anode formulations as depicted in this work and that used for large scale applications, lies essentially in the method of deposition of Ni. NiO films are expected to form upon impregnation and calcination, whereby they cover the entire surface area of the porous anode, including the interface between the electrolyte and the anode. After reduction of NiO to Ni, porosity is further induced, which results in the requisite 3 phase boundary length at the electrode-electrolyte interface. The fundamental principles of electrochemical reactions occurring at the interface remain the same, but the quantity of Ni being much lesser now, electrical conductivity over the bulk of the anode may be lessened – this problem must be avoided. Therefore, sufficient Ni will be needed to provide adequate electrical conductivity and yet much lesser than 30-35 vol % as estimated by the percolation model. The optimal doping of Ni in a pre-existing microstructure becomes an important parameter. Yet another important parameter that emerges in this paper is, the role of morphology. Not only does electrical conductivity depend upon the porosity and tortuosity but so do the electrochemical reactions at the 3 phase boundary. The role of graphite and polystyrene (both pore formers but with diverse particle morphologies) on cell performance is discussed in this paper.

EXPERIMENTAL

Tape casting technique was used to fabricate porous slabs of YSZ for conductivity testing, as were bi-layers (anode and electrolyte) for 'button' cells. Characterization by SEM, pore size analysis, particle size distribution measurements and electrical conductivity measurements have all been described in the following section.

Materials

Yttria doped Zirconia powder (TZ-8YS, Tosoh Corporation, Japan) was used as base powder for making anode and electrolyte. The mean particle size and surface area were 90 nm and 5.57 m^2/g respectively. Graphite (325 mesh, Alfa Aesar) and Polystyrene (Acros Organics) were used as pore formers. Ethanol (99.9%, Merck) and Methyl Ethyl Ketone (99%, Merck) were used as solvents. Oleic Acid (CDH, India) was used as dispersant. Polyvinyl Butyral (Chempure Laboratories, Mumbai) was used as binder and Polyethylene Glycol (Qualigens) was used as plasticizer.

Fabrication

SOFC 'button' cells were fabricated by tape casting the electrolyte layer first, on a glass sheet. Multiple anode layers (4 to 6) were cast on top of the electrolyte to achieve the required thickness. The pore former used in the anode was either graphite or polystyrene. Circular discs of diameter 14.7 mm were cut from these tapes and sintered to 1450°C for 10 hr. The graphite or the polystyrene pore formers burn off, leaving behind pores of different shapes and sizes. The cathode was prepared by brush painting a slurry containing $La_{0.75}Sr_{0.25}MnO_3$ (LSM), YSZ and graphite in 40:40:20 ratio. LSM was prepared in the lab by drip pyrolysis[9]. After painting the cathode, the cell was sintered at $1250°C$ for 2 hr. The active cell area is 0.5 cm^2.

Rectangular slabs for conductivity studies were also fabricated by tape casting. Tape casting was preferred over other techniques like isostatic cold pressing, in order to simulate the microstructural conditions in the button cell. Multiple anode layers were cast on top of a glass sheet, so that a final sintered thickness of about 2 mm is achieved. The green tapes were cut into slabs and sintered to 1450 °C for 10 hr. The final sintered slab dimensions are ca. $20×7×2$ mm. Slabs were used for the 4 point probe conductivity experiments. The fabrication procedure has been explained extensively in an earlier paper[8].

Nickel Deposition

Nickel film was deposited into the porous anode by impregnation of Ni-salt solution. A 5M aqueous solution of $Ni(NO_3)_2$ was used. The porous electrode slabs for conductivity measurements were impregnated by dipping the slabs inside the solution (solution excess). The porous anode of the SOFC was impregnated by dropping the solution on to anode (incipient wetness). The impregnated samples were then calcined at $350°C$ for 4 hr to decompose the $Ni(NO_3)_2$ to NiO. The impregnation step was repeated to achieve desired loadings. The NiO loaded electrode slabs and SOFC were connected to the electrical conductivity setup and I-V characterization setup respectively and were exposed to flowing H_2 gas. The NiO reduced to Ni at temperatures less than 450 °C.

Characterization
Particle Sizes, Porosity and Morphology

The particle size distributions of the starting powders were analyzed using laser particle size analyzer (Ankersmid - C50) and Nano-Sizer (Malvern). The porosity of the anodes was analyzed by water uptake method by measuring the mass of water adsorbed by the sample. The porosity of a few samples was measured by Hg porosimetry (Quantachrome – Poremaster) and the values were consistent with the ones calculated by water uptake method. So water uptake method has been used as a fast and suitable method for measuring the porosity of the samples. The pore size distributions of the sintered samples were measured by Hg intrusion porosimetry.

The microstucture of the starting materials and the sintered samples were analyzed using Scanning Electron Microscopy (SEM) (Zeiss EVO 50). The thickness of the electrolyte and electrodes were also measured by SEM.

Conductivity

The electrical conductivity of Ni-loaded slabs were measured by 4 probe DC conductivity method. AUTOLAB – PGSTAT30 was used as the constant current source. Four Ag wires were connected to the slabs using conductive Ag ink (Alfa Aesar). Measurements were done in a reducing atmosphere, with a H_2 flow rate of 60 ml/min upto a temperature of $700°C$.

I-V Curves

The surface of the anode and cathode of the SOFC were coated with a thin layer of Ag ink. Ag wires were attached to the anode and cathode with Ag ink. The anode side of the SOFC was then attached to a ceramic tube and sealed using ceramic paste (Aremco 552 VFG). H_2 gas was passed on the anode side at a flow rate of 60 ml/min and the cathode was exposed to air. The whole set up was kept inside a split-furnace. I-V characteristics were measured using an electrical load which was assembled in-house.

RESULTS AND DISCUSSION

Performance of SOFC's with Anodes made from Graphite Pore Formers
Figure 1 shows that the performance of the fuel cell increases as temperature increases. The data presented here are for a cell made using graphite as pore former having an initial porosity of 50.5 % and having a Ni content of 9.4 vol.%. The peak power densities at the three temperatures, 650°C, 750°C and 850°C were 10.5 mW/cm^2, 30 mW/cm^2 and 51 mW/cm^2 respectively. These values are much lower than those obtained in literature with similar electrolyte thickness. Initially the low performances were attributed to sintering of Ni which might very likely occur, as the Tammann temperature of Ni is 591°C and sharp conductivity losses were observed at low Ni doping levels in earlier work (Singh et al.)[8]. In a bid to demonstrate this further, cell performances over long periods of time were investigated.

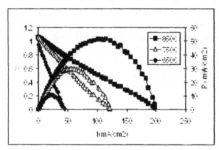
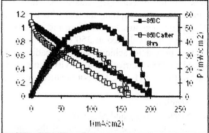

Figure 1. The effect of temperature on V-I curves for cell with Ni impregnated anode

Figure 2. Loss of performance with time for cell with graphite based anode when operating

Figure 2 shows the performance of the same cell mentioned above, as a function of time at 850°C. The peak power decreased from 51 mW/cm^2 to 35 mW/cm^2 after 8 hr. Performance curves reveal an increase in activation polarization, but no appreciable increase in the Ohmic polarization, as evidenced by a constant slope in both the I-V curves. Conductivity tests done on anodes with ca. 9 vol% Ni and porosity 50% reveal that the conductivity of the samples are stable for a long duration[8]. Sintering has no appreciable effect on the overall conductivities of the porous Ni-YSZ matrix, at Ni loadings of around 9 vol%, where as it does have a profound effect at lower Ni loadings (4-5 vol.%) as shown in our previous work[8]. But sintering could affect the Three Phase Boundary (TPB) region leading to lesser Three Phase Boundary Length (TPBL). This could explain the increase in activation polarization, without significantly impacting Ohmic polarization.

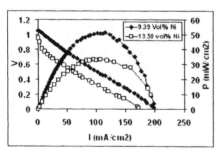

Figure 3. Effect of increasing Ni content on performance of cells with Ni impregnated anodes prepared using graphite as pore former

The effect of increasing the Ni loading on graphite based cells was studied. The performance was expected to increase with Ni loadings. But the

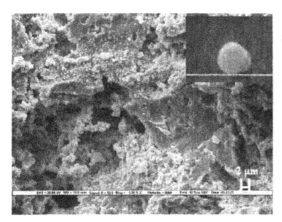

Figure 4. SEM of 9.39 vol.% Ni anode, graphite pore former – 50.5 % porosity. Ni particle size (inset) is 500nm.

results were counter intuitive. Figure 3 shows the I-V curves for a cell with 9.4 vol.% Ni and for a cell with 13.6 vol.% Ni, with both anodes having an initial porosity of 50%. The performance of the cell with the higher Ni loading is lower than the other cell. The I-V curves show that with increase in Ni loading, there is an increase in the activation polarization, but as expected there is a slight decrease in the Ohmic polarization. The slight decrease in Ohmic polarization is expected as the conductivity of the bulk increases with increase in Ni loadings but the advent of activation polarization with increase in loading suggests another major effect, i.e., role of Ni particle size vis-à-vis the microstructure. Microstructure studies were done on these cells in order to find out the reasons for the increase in activation polarization with increase in Ni. Figure 4 shows the SEM of the anode of a cell with 9.4% Ni loading. The small white spheres are the Ni particles. The inset shows a single Ni particle. The average size of a Ni particle is around 500 nm while the majority graphite based pores are less than 2 μm in size. Hence a few Ni particles are sufficient to block some pores. Hence as we increase the Ni loading, pores may become inaccessible too and hence Three Phase Boundary Length is likely to decrease, thereby inducing activation polarization. These studies reveal that the size and shape of the pores play a very important role in the cell performance.

Performance of SOFC's with Anodes made from Polystyrene Pore Formers

Cells made using polystyrene as the pore former in the anode, have spherical pores and an average pore size of around 20 μm. Figure 5 shows the effect of temperature on cell performance. The data shown here is for a cell made with polystyrene pore former and Ni loading of 10.6 vol.%, with the

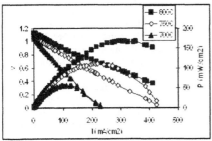

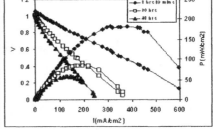

Figure 5. The effect of temperature on V-I curves for cell with Ni impregnated anode

Figure 6. Loss of performance with time for cell with polystyrene based anode when

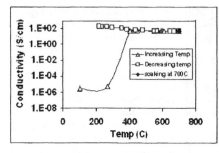

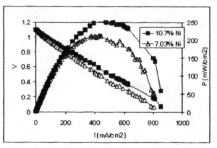

Figure 7. Conductivity of an anode made using polystyrene as pore former, with Ni - 8.8 vol.% as a function of time and temperature.

Figure 8. Effect of increasing Ni content on performance of cells with Ni impregnated anodes prepared using polystyrene as pore former

anode having an initial porosity of 57%. Cell performance as measured by peak power densities at the three temperatures, 700°C, 750°C and 800°C were 55 mW/cm², 110 mW/cm² and 168 mW/cm² respectively. Polystyrene based anode microstructure appears to be very beneficial for cell performance. In either case, the electrolyte thickness remained the same, about 20 µm. However, in order to assess the damage due to sintering of Ni as in the previous case, cell performances over long durations of time were investigated.

Figure 6 shows the performance of the same cell mentioned above as a function of time at 800°C. The maximum power density of 182 mW/cm² dropped to 85 mW/cm² after 30 hr and further dropped to 58 mW/cm² after 48 hr of continuous operation. The I-V curves show an increase in activation polarization as well as ohmic polarization.

Conductivity studies were done on polystyrene based anodes with different loadings of Ni. Figure 7 shows the conductivity of an anode with a Ni loading of 8.8 vol.% and an initial porosity of 56 %, as a function of time. The anode was soaked at 700°C for 6 hr and there was no significant

reduction in conductivity in these conditions. Therefore, the sintering does not seem to have a significant detrimental effect on the overall conductivity at Ni loadings around 8.8 vol.% and higher. Hence, the decrease in performance could entirely be due to the effect of sintering phenomena at the TPB region. Higher Ni particle size will lead to decrease in the TPBL and decrease in electrical connectivity in the TPB region.

The effect of increasing the Ni loading on polystyrene based cells was studied. Figure 8 shows that in this case the cell performance increases on increasing the Ni loading. The maximum power density increased from 213 mW/cm² at 7 vol.% Ni to 250

Figure 9. SEM of 10.7 vol.% Ni anode, polystyrene pore former – 57 % porosity. Ni particle size (inset) is > 1µm.

mW/cm^2 at 10.7 vol.% Ni. The initial porosity of both the cells were 57 vol.%. The increase in performance could be both due to an increase in conductivity as well as an increase in TPBL as the Ni loading in increased. In contrast to the graphite based cell, the pore sizes in polystyrene based cells are around 20 μm and the Ni particle are around 1 μm as shown in Figure 9. The Ni particles do not block the pore in this case and the performance is shown to increase with Ni loading.

CONCLUSIONS

SOFC 'button' cells with low Ni content have been fabricated, where Ni was incorporated into a preformed porous matrix by aqueous impregnation technique. Different porous structures were introduced in the anode by using different pore formers like graphite or polystyrene. It is apparent from the results in this paper, that there are major differences in cell performance and in performance loss, with variation in pore morphology. Large pore sizes appear to provide better performance as Ni particles are less likely to block pore openings and pathways for gas transport. However smaller pore sizes as generated by graphite pore former, while tending to lower TPBL, also stabilize the Ni particles by checking the Ni particle agglomeration due to sintering – this may purely be the effect of pore size constraints. Sintering of Ni particles under SOFC operating conditions is inevitable and is responsible for loss of connectivity among Ni particles thereby leading to extra Ohmic losses and/or losses in TPBL. The study reveals that proper optimization of the pore sizes and shape is necessary for improving performance in cells with low Ni content. This is part of an ongoing study.

REFERENCES

1. N. Q. Minh, *J. Am. Ceram. Soc.*, **76** (1993) p563.
2. Nishant M. Tikekar, Tad J. Armstrong, and Anil V. Virkar, *J. Electrochem. Soc.* **153(4)** (2006), A654.
3. D. Waldbillig, A. Wood and D. G. Ivey, *J. Electrochem. Soc.* **154(2)** (2007), B133
4. R. M. C. Clemmer, and S. F. Corbin, *Solid State Ionics*, **166** (2004), p251.
5. Swadesh K Pratihar, A. Das Sharma, andH. S. Maiti, *Materials Research Bulletin* **40** (2005) p1936.
6. S. Park, J. M. Vohs, and R. J. Gorte, *Nature*, **404** (2000) p265.
7. Y. Huang, J. M. Vohs, and R. J. Gorte, *Electrochem. Solid-State Lett.*, **9(5)** (2006), A237.
8. C. Anand Singh and Venkatesan V. Krishnan, *ECS Transactions - Chicago* Volume 6, "Design of Electrode Structures", 2007 (In Press).
9. Robert J. Bell, Graeme J. Millar, and John Drennan, *Solid State Ionics*, **131** (3-4) (2000), p211.

THE REDUCTION OF NiO-YSZ ANODE PRECURSOR AND ITS EFFECT ON THE MICROSTRUCTURE AND ELASTIC PROPERTIES AT AMBIENT AND ELEVATED TEMPERATURES

Thangamani Nithyanantham*, Saraswathi Nambiappan Thangavel, Somnath Biswas , and Sukumar Bandopadhyay
Institute of Northern Engineering, University of Alaska, Fairbanks, AK 99775, USA

ABSTRACT
NiO-8YSZ anode supported half cells were reduced for different periods of time at 800°C in 5% H_2 atmosphere. The reduction of NiO is associated with significant changes in the crystalline phases, porosity and microstructure and hence in elastic properties. The Young's modulus was determined after reduction at room temperature and at elevated temperatures using Impulse Excitation Technique. The Young's modulus was found to be decreasing with increasing porosity. The decrease in the Young's modulus was about 50% and can be attributed mainly to the changes in the microstructure particularly increasing porosity from about 12% to 37% after 8 hrs of reduction. The elastic properties of the reduced anodes were estimated at elevated temperatures in air.

INTRODUCTION
Recently, there have been tremendous research interests on the development of solid oxide fuel cells (SOFCs) due to its potential as a highly efficient source of environmentally clean energy.[1-8] SOFCs, which convert the chemical energy of a gaseous fuel into electricity, have a basic structure of a highly dense ion-conducting electrolyte layer, sandwiched between electron conducting anode and cathode layers.[2,3] Since the electrolyte is kept as thin as possible to decrease ohmic losses through it and to lower the operating temperature of the cell and the cathode is a weak and compliant material,[7,9-12] often in certain designs the anode as the mechanical support strongly determines the structural and mechanical integrity and reliability of the cell.[1,2,10]

In general, Ni-Y_2O_3 stabilized ZrO_2 (Ni-YSZ) based porous cermet is the most favored anode material in SOFCs, due to its enviable properties of excellent electro-catalytic activity for the H_2 oxidation reaction at the triple phase boundary (TPB) between Ni, YSZ and gaseous H_2,[13-15] high electronic conductivity,[6] high mechanical stability and compatibility with YSZ electrolyte,[16,17] and structural stability at high operating temperatures (700 to 1000°C) in H_2 atmosphere.[5,6] The stability and performance of Ni-YSZ anode structure is crucially dependent on the phase distribution and microstructure in the cermets. Usually, the Ni-YSZ anode structure is fabricated by reducing a NiO-YSZ anode precursor structure which readily generates the desired porosity, microstructure, and mechanical strength in the anode.[1,4-6] It is observed that the microstructure in the Ni-YSZ cermets is considerably dependent on the composition, morphology and process of synthesis of the precursor, resulting in significant changes in the overall performance of the SOFC.[6,7,18-20] For the optimum performance of the anode, at least 40 vol% of Ni [21,22] is required along with a homogeneous distribution of Ni and YSZ phase.[20] The distribution of the metal ions controls the coefficient of thermal expansion (CTE) of the anode and determines the structural stability of the SOFC during operation.[6,23] Although, small particle size is advantageous for obtaining high length of triple phase boundary (TPB), however, bigger particle size helps to reduce the effect of Ni on the CTE and often results in a well connected YSZ structure, thereby improving the thermo/electro-mechanical performance of the anode.[7,20] However, the themomechanical properties, particularly the elastic properties are sensitive to porosity and the decrease of Young's and shear moduli of NiO-YSZ with porosity have been reported earlier by Seluk and Atkinson[24] and Radovic and Lara-Curzia.[4,25] The reported data on the elastic properties are determined mostly at the ambient temperature.

In this paper, we report a systematic analysis of the phase formation and development of cermet microstructure in the anodes derived by reducing NiO-8YSZ anode precursor supported half cells with dense electrolyte on one side for selected time periods. The mechanical properties of the reduced samples were studied at room temperature as a function of porosity and Young's moduli of the as-sintered and reduced anodes were evaluated as a function of temperature in air. The results were correlated with the initial porosity, composition and oxidation of Ni at the elevated temperatures.

EXPERIMENTAL PROCEDURE
The half-cell samples used in this investigation were prepared by a proprietary process involving tape casting of anode precursor, spray coating of electrolyte and sintering. The square laminates have a dense, about 8 μm thick electrolyte layer of 8YSZ (stabilized by 8 mol% Y_2O_3) supported by a about 600 μm thick NiO-8YSZ (80:20 vol%) anode precursor layer. The as-sintered half cells were cut into rectangular bars (50 x 30 x 0.6mm) and reduced in a gas mixture of 5% H_2 – 95% Ar at 800°C. The reduction reaction was carried out in an autoclave set-up, where the samples are kept at the above temperature for 1 h prior to the reduction to attain a thermal equilibrium before the reaction starts. Then the gas mixture was introduced in the chamber at a constant flow rate of 1 SLPM and the temperature-chemical environment condition was maintained for selected time periods of 10 min, 30 min, 2 h and 8 h. After that the samples were cooled down to room temperature without changing the reducing gaseous environment.

The porosity and density values in the as received and the reduced half-cell samples were estimated by water immersion method as given in ASTM Standard C20-00. Since the electrolyte layer is very thin in comparison to the anode, these values are a good approximation of the true values of porosity/density in the anode. The fraction of NiO reduced in the samples was determined by thermogravimetric analysis (TGA) carried out in air. The crystal structure of the reduced samples was analyzed with x-ray diffraction (XRD) studies, using a Rigaku diffractometer with 0.15418 nm Cu K_α radiation. The microstructures were studied by a JEOL JSM-7000 scanning electron microscope (SEM). An accelerating voltage of 10 kV was used to resolve the images of the Ni-YSZ cermet structure. Hardness of the samples was measured by Vicker's indentation method on the anode surface with a load of 500g for 15s. Room temperature elastic properties (Young's and shear moduli) of the as-received, partially and fully reduced samples were determined by the impulse excitation technique using the commercially available Buzz-o-sonic nondestructive testing system (BuzzMac International, Glendale, WI) which measures the fundamental vibration frequencies. Young's moduli of the half cells were evaluated at elevated temperatures using a custom designed high temperature module of the testing system. A cylindrical alumina base was used as a stage on which the test specimen (50x15x0.6mm) was suspended on a thin wire support. A computer controlled impulse tool which fired an alumina rod onto the sample, was used to give mechanical impulse at the bottom of the rectangular bar and the vibration frequencies were transferred to a microphone through a sound guide. Delivering mechanical impulse in a definite time interval (1 min) and data acquisition was managed by the Buzz-o-sonic testing system. The whole experimental setup was housed inside a furnace and the temperature was controlled by a programmable controller..

RESULTS AND DISCUSSION
Evaluation of Porosity, Crystalline Phase and Microstructure in Ni-8YSZ Anode
As shown in Fig. 1, the relative porosity value in the reduced samples increases with the reduction time, i.e., with the fraction of reduced NiO. The porosity has increased from a value of 12% in the as received samples to 36.7% in the 8 h reduced samples. This type of behavior is as expected, since, the specific volume of metallic Ni is significantly smaller than that of NiO. Since the YSZ has

little effect on the reduction process at that temperature ($\leq 1000°C$) in H_2 atmosphere and the change in mass is attributed only to the change of NiO to Ni.[4,7,25]

The high temperature sintering (generally, $1100°$ to $1500°C$) of the green anode precursor structure after its casting results in essential densification and neck formation between the particles in the anode, which is necessary for the formation of percolation paths.[1,6] About 1 mol% NiO acts as a sintering aid for YSZ.[26] The solubility of NiO in YSZ depends on both the sintering temperature and Y_2O_3 content. A maximum solubility of 2.5 mol% in 8YSZ is reported at $1200°C$.[26,27] The NiO phase which contracts ~ 25 vol% after the reduction to Ni, does not wet YSZ very well with a wetting angle of ~ $120°$.[28] Moreover, Ni has high surface mobility at higher temperature with a strong propensity to agglomerate.[29] As a result, the performance and durability of the anode highly depends on the sintering of a well-packed green structure to obtain a rigid YSZ network capable of restricting the Ni phase from agglomeration.

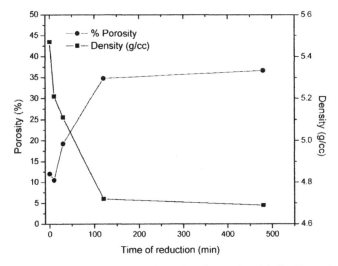

Fig. 1. Measured values of porosity and density of the reduced half-cell samples as a function of reduction time.

Since the half cells have an electrolyte layer, it is difficult to estimate the remaining mass of NiO in the partially reduced anodes from the weight of the test samples. Hence, thermal analysis (Fig. 2) was carried out in air in order to determine the fraction of reduced NiO in the half cell samples reduced for various time periods. Thermogravimetric analysis of the reduced samples revealed that the oxidation starts at about $400°C$ and completes before $850°C$ depending upon the Ni content in the cermets. From the weight gain due to the oxidation of the reduced Ni in the samples, the initial fractions of reduced NiO (%) in the samples were calculated as shown in Fig. 3. It appears that the first few hours in reducing the NiO in the anode are decisive as 2 h of exposure to the reducing condition ($800°C$ in 5% H_2 environment) is enough to reduce ~ 80% of NiO. Further exposure to the reducing conditions could only cause a very marginal increase in the reduced amount of NiO. The amount of reduced NiO after 8 h of exposure to the reducing conditions is only 81.3%. The inset figure compares the x-ray diffractograms in the anode surface of the as sintered half cells with that of the samples reduced for 10 min, 30 min, 2 h and 8 h, respectively.

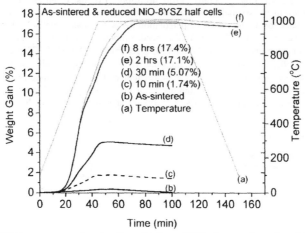

Fig. 2. TG thermograms, showing the oxidation of Ni in the samples reduced for (b) as-sintered (c) 10 min, (d) 30 min, (e) 2 hrs and (f) 8 hrs when subjected to a temperature profile (a) in air.

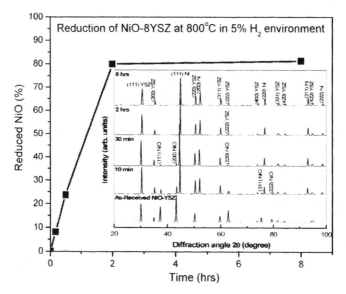

Fig. 3. The fraction of reduced NiO (%) in the samples as a function of reduction time. The inset figure shows the x-ray diffractograms in the anode surface of the as received and the reduced samples.

The as sintered sample shows strong peaks of NiO and relatively weaker peaks of cubic-ZrO_2. On reduction, the NiO transforms to cubic Ni. The NiO peaks gradually diminish with the increase in reduction time in the samples reduced for 10 or 30 minutes, however, the lattice parameter of Ni remains more or less unchanged with reduction time. Although, as par TGA measurements a part of NiO remains unreduced in the samples even after reduction for 8 h, no NiO peaks were observed in their x-ray diffractograms. Furthermore, a very slow scan XRD study (0.01° step size, 2 s dwell time) after pulverizing the half cells leaves no trace of NiO in the samples reduced for ≥ 2 h or any Ni in the as-received samples. The percentage of reduced NiO was also calculated using theoretical final mass change upon reduction and initial mass of NiO in the sample.[4] The results were compared with the data used in Fig. 3 and no significant differences were noticed. A reduction behavior of similar trend was reported by Radovic and Lara-Curzia[4] for 75 mol% NiO-YSZ anodes in the 4% H_2 atmosphere. In Fig. 4, the SEM micrographs illustrate a systematic study of the effect of reduction on the microstructure development in forming the Ni-8YSZ anode.

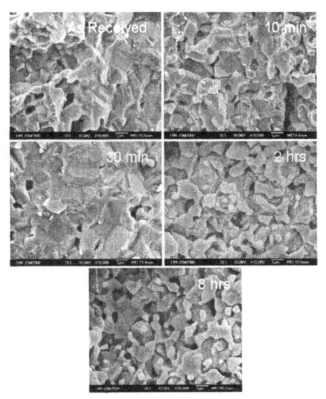

Fig. 4. SEM micrographs of the anode surface of (a) as-received, (b) 10 min, (c) 30 min, (d) 2 h and (e) 8 h reduced samples.

The as received NiO-8YSZ anode precursor (Fig. 4a) is relatively dense with very few pores which are neither uniformly distributed nor have regular shapes. The exposure to the reducing conditions for 10 min or 30 min has formed a thin layer of Ni on the NiO particles. Further exposure to the reducing conditions has reduced the volume of NiO considerably (not reflected in XRD) which leads to the development of a typical porous cermet microstructure of homogeneous dispersion of Ni particles in 8YSZ with interconnected pores (Fig. 4d & e). It is also observed that the pores that were formed due to shrinkage of NiO particles during their reduction into Ni are much bigger than those initially present in NiO-8YSZ samples.

Mechanical Properties of Ni-8YSZ Anodes at Room Temperature

The formation of the desired porous microstructure of Ni-8YSZ is very crucial for the development of successful anode systems with high amount of TPB contacts.[6,7] High electrical power is obtained through high electrochemical conversion rates for the H_2 oxidation reaction at the anode leading to high current densities at low over potentials. High conversion rate requires high electro-catalytic activity in anode which is directly dependent on the quantity of TPB contacts. Small particles and pore sizes favors in obtaining high TPB. However, the mechanical stability of the anode structure at high operating temperature is essentially dependent on the interconnected YSZ network.

Fig. 5 shows the Vicker's hardness values of as-received and the reduced anode samples plotted as a function of porosity. The as-received precursor sample which has higher density (5.47 g/cc) and low porosity (12%) shows a hardness value of 5.5 GPa. After reduction, the hardness value has reduced to less than 1 GPa in the 8 h reduced samples with an increased porosity of 36.7%. The increase in porosity as a result of the reduction has severely affected the hardness of the anode. It is interesting to note that the scattering in the hardness values also decreases with increasing porosity. Even though they are highly porous, the 2 h or 8 h reduced anode samples have negligible scattering in their hardness values in comparison to the as received samples. This type of behavior can be attributed to the higher metal content present in the highly reduced porous anodes.

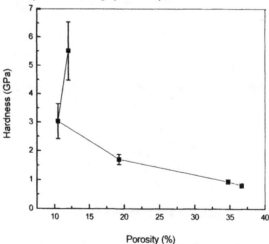

Fig. 5. Hardness values plotted as a function of porosity in the as received and the reduced anode samples.

In Fig. 6, the Young's (E) and shear (G) moduli of as-received and the reduced anodes, characterized with the impulse excitation technique at room temperature, are plotted as a function of volume fraction of porosity in the samples. The measured values of Young's and shear moduli in the as-sintered NiO-8YSZ samples are found to be 131.3 and 54.5 GPa, respectively. The elastic moduli values decrease significantly with the increase in the volume fraction of porosity. The observed decrease in elastic moduli is evidently a result of changes in the composition of the anode and/or increase in porosity.[4,7] Since, the elastic moduli of fully dense Ni ($E = 200$ GPa and $G = 77$ GPa),[30] NiO ($E = 220$ GPa and $G = 84$ GPa)[30] and 8YSZ ($E = 220$ GPa and $G = 83.3$ GPa)[24] are comparable to each other, the changes in the chemical composition of the anode after reduction is expected to have trivial effects on the magnitude of the effective elastic moduli of the anode material. Therefore, the observed decrease of ~ 44% in the Young's modulus (~ 40% in shear modulus) after 8 h of reduction of the anode samples is predominantly due to the significant increase in porosity in the samples.

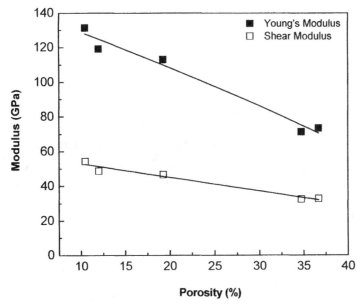

Fig. 6. Room temperature Young's and shear moduli values plotted as a function of porosity in the as sintered and the reduced anode samples.

Selcuk and Atkinson[24] reported similar type of porosity dependence of elastic moduli at room temperature in 75 mol% NiO-YSZ anode samples with as much as 14% initial porosity. They found that the Young's modulus of fully dense NiO-YSZ (205-218 GPa) decreases drastically with the increase in the volume fraction of porosity. Analogous results were also obtained by Radovic and Lara-Curzio[4] in 75 mol% NiO-YSZ samples (23% porosity) with a lower value of E and G in the range of 93.6-103.3 GPa and 36.2-40.3 GPa, respectively. Based on their proposed empirical model, they predicted that the difference in the elastic moduli between hypothetical fully dense NiO-YSZ and Ni-

YSZ is about 4%. Therefore, the observed large decrease in elastic moduli is primarily due to the significant increase in porosity with the reduction of NiO.

The changes in hardness and elastic moduli after reduction in the Ni-based anode materials results in a re-distribution of residual stresses in the SOFC components. A detailed analysis of the effects in correlation with microstructure is necessary in order to predict the conditions for optimal performance and reliability of the SOFCs.

Young's Moduli of Ni-8YSZ Anodes at Elevated Temperatures

The temperature dependent elastic properties of the NiO-YSZ and reduced anodes were evaluated in air. The Young's moduli of the anodes are dependent on the initial porosity, composition and the oxidation of Ni in air. The porosities of the as-sintered and reduced half cells have profound effect on their room temperature as well as high temperature Young's moduli. In addition to this, the onset of oxidation in cermets at elevated temperatures influences the modulus significantly. Along with the other material properties like density, the coefficient of thermal expansion (TEC) of the anode is also essential to determine the modulus as a function of temperature. A TEC value of $14.4 \times 10^{-6} K^{-1}$ was determined at 1000°C in air for the as-sintered half cell, which was used to calculate the modulus at the elevated temperatures. Fig. 7 shows the Young's moduli of the as-sintered and reduced anodes as a function of temperature in air. The temperature dependence of Young's modulus in air can be discussed on the basis of the composition of the starting half cell.

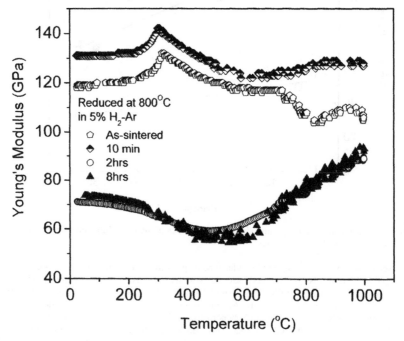

Fig. 6. Young's moduli values plotted as a function of temperature for the as-sintered and the reduced anode samples.

The half cells which have NiO as one of the constituents show a distinctively separate elastic behavior through out the temperature profile from the cermets which have no or negligible amount of NiO as one of their constituent phases. At the outset, the NiO containing half cells have significantly higher moduli at room temperature since they have less porosity than the fully reduced cermets. As temperature progresses, an increase of the moduli is observed from 200°C which reaches a maximum value at around 300°C, prior to a sharp decline as the temperature progresses further. This well defined peak in the modulus vs. temperature curve at 300°C can be related to the structural transition of NiO from rhombohedral to cubic at this temperature. This structural transition affects many material properties of NiO including thermal expansion[31]. No such structural transition assisted change in the modulus curve is noticed in the other cermets that have no or negligible amounts of NiO in their composition. The decreasing moduli of the NiO rich half cells reach a steady state at 500°C and remain unchanged till 1000°C. The reason for the observed decline in the modulus values of the as-sintered half cell after 750°C is not clear yet. Up to 500°C the reduced cermets exhibit a decreasing trend in their moduli and then an increase is noticed.   This rise in the modulus values can be correlated to the oxidation of the Ni in the air. The formation of NiO scales over the Ni grains decreases the porosity which in turn increases the modulus of the half cell.

CONCLUSIONS

Highly porous Ni-8YSZ anodes for SOFC applications have been developed by reducing a NiO-8YSZ anode precursor structure in a gas mixture of 5% H_2-95% Ar at 800°C for selected time periods. XRD and SEM analysis in the reduced samples disclose the formation of the Ni-8YSZ cermet structure with desired porosity and microstructure. The porosity in the anode samples, which increases with the increase at the fraction of reduced NiO, severely affects the hardness and elastic moduli of the anode samples at room temperature. It is found that a hardness value of 5.5 GPa in the as-sintered anode samples (12% porosity) reduces to less than 1 GPa in the 8 h reduced samples (36.68 % porosity). Similarly, a decrease of ~ 44% in the Young's modulus and ~ 40% in shear modulus is observed in the 8 h reduced samples in comparison to the as-sintered anode precursor. Young's moduli of the as-sintered and reduced anodes were evaluated as a function of temperature in air. The results were correlated to the initial porosity, composition and oxidation of Ni at the elevated temperatures.

ACKNOWLEDGEMENT

This work has been carried out with financial support from the United States Department of Energy project grant # DE-FG36-05GO15194. The authors sincerely thank Materials Research and System Inc., Salt Lake City, UT, for providing the samples.

REFERENCES

1.   Primdahl, S., Sorensen, B. F. and Mogensen, M., Effect of nickel oxide/yttria-stabilized zirconia anode precursor sintering temperature on the properties of solid oxide fuel cells. *J. Am. Ceram. Soc.*, 2000, **83**, 489-494.
2.   Singhal, S. C., Solid oxide fuel cells for stationary, mobile, and military applications. *Solid State Ionics*, 2002, **152-153**, 405-410.
3.   Haile, S. M., Fuel cell materials and components. *Acta Mater.*, 2003, **51**, 5981-6000.
4.   Radovic, M. and Lara-Curzio, E., Elastic properties of nickel-based anodes for solid oxide fuel cells as a function of the fraction of reduced NiO. *J. Am. Ceram. Soc.*, 2004, **87**, 2242-2246.
5.   Ramanathan, S., Krishnakumar, K. P., De, P. K. and Banerjee, S., Powder dispersion and aqueous tape casting of YSZ-NiO composite. *J. Mater. Sci.*, 2004, **39**, 3339-3344.
6.   Jiang, S. P. and Chan, S. H., Development of Ni/Y_2O_3-ZrO_2 cermet anodes for solid oxide fuel cells. *Mater. Sci. Tech.*, 2004, **20**, 1109-1118.

7. Wang, Y., Walter, M. E., Sabolsky, K. and Seabaugh, M. M., Effects of powder sizes and reduction parameters on the strength of Ni-YSZ anodes. *Solid State Ionics*, 2006, **177**, 1517-1527.

8. Delaforce, P. M., Yeomans, J. A., Filkin, N. C., Wright, G. J. and Thomson, R. C., Effect of NiO on the phase stability and microstructure of yttria-stabilized zirconia. *J. Am. Ceram. Soc.*, 2007, **90**, 918-924.

9. Atkinson, A. and Selcuk, A., Mechanical behaviour of ceramic oxygen ion-conducting membranes. *Solid State Ionics*, 2000, **134**, 59-66.

10. Selcuk, A., Merere, G. and Atkinson, A., The influence of electrodes on the strength of planar zirconia solid oxide fuel cells. *J. Mater. Sci.*, 2001, **36**, 1173-1182.

11. Dokiya, M., SOFC system and technology. *Solid State Ionics*, 2002, **152/153**, 383-392.

12. Gaudon, M., Menzler, N. H., Djurado, E. and Buchkremer, H. P., YSZ electrolyte of anode-supported SOFCs prepared from sub micron YSZ powders. *J. Mater. Sci.*, 2005, **40**, 3735-3743.

13. Setoguchi, T., Okamoto, K., Eguchi, K. and Arai, H., Effects of anode material and fuel on anodic reaction of solid oxide fuel cells. *J. Electrochem. Soc.*, 1992, **139**, 2875-2880.

14. Jiang, S. P. and Badwal, S. P. S., Hydrogen oxidation at the nickel and platinum electrodes on yttria-tetragonal zirconia electrolyte. *J. Electrochem. Soc.*, 1997, **144**, 3777-3784.

15. Kim, S. J., Lee, W., Lee, W. J., Park, S. D. and Song, J. S., Preparation of nanocrystalline nickel oxide-yttria-stabilized zirconia composite powder by solution combustion with ignition of glycine fuel. *J. Mater. Res.*, 2001, **16**, 3621-3627.

16. Tsoga, A., Naomidis, A. and Nikolopoulos, P., Wettability and interfacial reactions in the systems Ni/YSZ and Ni/Ti-TiO2/YSZ. *Acta Mater.*, 1996, **44**, 3679-3692.

17. De Boer, B., Gonzalez, M., Bouwmeester, H. J. M. and Verweij, H., The effect of the presence of fine YSZ particles on the performance of porous nickel electrodes. *Solid State Ionics*, 2000, **127**, 269-276.

18. Brown, M., Primdahl, S. and Mogensen, M., Structure/performance relations for Ni/yttria-stabilized zirconia anodes for solid oxide fuel cells. *J. Electrochem. Soc.*, 2000, **147**, 475-485.

19. Lee, J. H., Heo, J. W., Lee, D. S., Kim, J., Kim, G. H., Lee, H. W., Song, H. S. and Moon, J. H., The impact of anode microstructure on the power generating characteristics of SOFC. *Solid State Ionics*, 2003, **158**, 225-232.

20. Zhu, W. Z. and Deevi, S. C., A review on the status of anode materials for solid oxide fuel cells. *Mater. Sci. Eng. A Struct. Mater. : Prop. Microstruct. Proces.*, 2003, **362**, 228-239.

21. Marinek, M., Zupan, K. and Macek, J., Preparation of Ni-YSZ composite materials for solid oxide fuel cell anodes by the gel-precipitation method. *J. Power Sources*, 2000, **86**, 383-389.

22. Lee, J. H., Moon, H., Lee, H. W., Kim, J., Kim, J. D. and Yoon, K. H., Quantitative analysis of microstructure and its related electrical property of SOFC anode, Ni–YSZ cermet. *Solid State Ionics*, 2002, **148**, 15-26.

23. Mori, M., Yamomoto, T., Itoh, H., Inaba, H. and Tagawa, H., Thermal expansion of nickel-zirconia anodes in solid oxide fuel cells during fabrication and operation. *J. Electrochem. Soc.*, 1998, **145**, 1374-1381.

24. Selcuk, A. and Atkinson, A., Elastic properties of ceramic oxides used in solid oxide fuel cells (SOFC). *J. Eur. Ceram. Soc.*, 1997, **17**, 1523-1532.

25. Radovic, M. and Lara-Curzio, E., Mechanical properties of tape cast nickel-based anode materials for solid oxide fuel cells before and after reduction in hydrogen. *Acta Mater.*, 2004, **52**, 5747-5756.

26. Linderoth, S. and Kuzjukevics, A., NiO in yttria-doped zirconia; pp. 1076-1085 in *Solid Oxide Fuel Cells V*, Proceedings of the Fifth International Synposium on Solid Oxide Fuel Cells

(Aachen, Germany, June 1997), Vol. PV 97-40. Edited by U. Stimming, S. C. Singhal, H. Tagawa and W. Lehnert. The Electrochemical Society, Pennington, NJ, 1997.

27. Kuzjukevics, A., Linderoth, S. and Grabis, J., Plasma produced ultrafine YSZ-NiO powders; pp. 319-330 in *High Temperature Electrochemistry : Ceramics and Metals*, Proceedings of the 17th Risø International Symposium on Materials Science (Roskilde, Denmark, Sep. 1996). Edited by F. W. Poulsen, N. Bonanos, S. Linderoth, M. Mogensen and B. Zachau-Christiansen. Risø National Laboratory, Roskilde, Denmark, 1996.

28. Nikolopoulos, P. and Sotiropoulou, D., Wettability between zirconia ceramics and the liquid metals copper, nickel, and cobalt. *J. Mater. Sci. Lett.*, 1996, **6**, 1429-30.

29. Murphy, M. M., Van herle, J., McEvoy, A. J. and Thampi, K. R., Electroless deposition of electrodes in solid-oxide fuel cells. *J. Electrochem. Soc.*, 1994, **141**, L94-L96.

30. Liu, C., Lebrun, J. L. and Huntz, A. M., Origin and development of residual stresses in Ni-NiO system : In-situ studies at high temperature by x-ray diffraction. *Mater. Sci. Eng. A*, 1993, **160**, 113-126.

31. Masashi, M., Tohru, Y., Hideaki, I., and Hiroaki, T., Thermal Expansion of Nickel-Zirconia Anodes in Solid Oxide Fuel Cells during Fabrication and Operation. J. Electochem. Soc. 1998, **145**, 1374-1381.

MICROSTRUCTURE ANALYSIS ON NETWORK-STRUCTURE FORMATION OF SOFC ANODE FROM NiO-SDC COMPOSITE PARTICLES PREPARED BY SPRAY PYROLYSIS TECHNIQUE

Hiroyuki Yoshida, Mitsunobu Kawano, Koji Hashino, and Toru Inagaki,
The Kansai Electric Power Company, Inc.
11-20 Nakoji 3-chome, Amagasaki
Hyogo 661-0974, Japan

Seiichi Suda and Koichi Kawahara
Japan Fine Ceramics Center
2-4-1 Mutsuno, Atsuta-ku
Nagoya 456-8587, Japan

Hiroshi Ijichi and Hideyuki Nagahara
Kanden Power-tech Co. Ltd.
2-1-1800, Benten 1-chome, Minato-ku
Osaka 552-0007, Japan

ABSTRACT
        The composite particles of NiO and samaria-doped ceria (SDC) having the core-shell structure and the dispersed structure were synthesized by spray pyrolysis method. The composite particles were sintered at various temperatures onto the electrolyte substrate, and the network-formation process was investigated by scanning electron microscopy (SEM), transmission electron microscopy (TEM), and the electrochemical characterization. In the case of the core-shell type particles, the SDC grains on the surface of the particles began to connect at the first step and the connection between the NiO grains in the neighboring particles started later. On the other hand, it was found that the connection between NiO grains started at almost the same time as that between SDC grains for the dispersed type particles. When the core-shell type composite particles were used, the connections among SDC grains generated and the grain size of NiO inside the particles grew at the initial stages of the sintering, and the densification of the anode was controlled. Therefore, the network-structure having wider connecting necks can be formed after high temperature sintering, and high-performance anode can be obtained. When the dispersed type composite particles are used, electronic conduction paths existed at the initial stages of the sintering since the connections among the NiO grains generated on the surface of the particles. Furthermore, large triple phase boundaries (TPBs) are obtained due to the well-dispersion of SDC and NiO grains, therefore, the particles have an advantage that high-performance can be obtained by sintering at lower temperatures.

INTRODUCTION
        Investigation of intermediate temperature Solid Oxide Fuel Cells (IT-SOFCs) operating at 600-800°C have been attracting more attention in recent years. The Kansai Electric Power Company, Inc. and Japan Fine Ceramics Center are also interested in IT-SOFCs. Polarization losses on anode and cathode as well as ohmic loss in electrolyte become higher as operating temperature decreases. The causes of cell voltage loss are lower ionic conductivity of electrolyte and lower electrochemical activities on anode and cathode at intermediate temperatures. It has been reported that doped lanthanum gallate compounds, La(Sr)Ga(Mg)O$_{3-\delta}$ (LSGM) give high ionic conductivity at intermediate temperatures[1,2], therefore, we selected LSGM as an electrolyte material. In order to commercialize IT-SOFC, not only the improvement of the electrolyte conductivity, but also the reduction of polarization loss at electrodes

at intermediate temperature should be accomplished.

Ni is used as an anode material since it is very active for adsorption of H_2 and has high electronic conductivity. Electrochemical reaction at electrode occurs at triple phase boundary (TPB), where ion, electron, and gas exist simultaneously. In order to improve the performance of the electrode, TPB should be increased. Therefore, ionic conductors such as electrolyte materials are generally composed with Ni. Among ionic conductors ceria-based anodes have been proposed as possible IT-SOFC anode materials since doped ceria compounds have high ionic and electronic conductivity under reducing atmosphere, and samaria-doped ceria (SDC) is considered to be the most attractive material among the ceria-based compounds[3]. Ni-SDC cermet anode gives high performance due to synergistic effect of enlarging reaction area and increasing paths for ionic and electronic conduction. Ni-SDC cermet is usually prepared by mixing and sintering SDC and NiO powders, which are larger than tenths of micrometer. If anode is composed of smaller SDC and Ni particles, electrochemical reaction area can be larger and anodic polarization can be lower than that of conventional anode. Therefore, we started developing the composite particles of NiO and SDC synthesized by spray pyrolysis technique.

Spray pyrolysis method has an advantage that composite particles with various internal structures can be individually synthesized, for example core-shell and dispersed structures. Furthermore, the anode fabricated from the composite particles synthesized by spray pyrolysis technique gave good performance at intermediate temperatures shown in the previous studies[4-8]. It was clarified by scanning electron microscopy (SEM) that the good performance of the anode manufactured from the composite particles was due to the proper network-structure between Ni and SDC after sintering[5,6,8]. In order to clarify the cause of the proper anode microstructure, the internal structure of NiO-SDC composite particles has been investigated by SEM and transmission electron microscope (TEM)[6-8]. However, it is not clear how the composite particles having the core-shell structure and the dispersed structure can form the proper network-structure of the anode. The relationship between the composite powders and the network structure of the anode has only preliminary investigated so far[6]. In this study, therefore, the network-formation processes from core-shell structure particles and dispersed structure particles were investigated in detail, and it was discussed what were the factors to obtain the high-performance anode from the composite particles with each internal structure.

EXPERIMENTAL
Spray pyrolysis synthesis

In this study, two types of NiO-SDC composite particles were synthesized by spray pyrolysis technique, the details of which were precisely described in the previous papers[4,8]. One is the particles with core-shell structure, in which SDC covered on the surface region of NiO core. Another one is the particles with dispersed structure, in which NiO and SDC primary particles having around 10 nm in diameter are dispersed. Atomic ratio of Ce:Sm was set for 80:20 to give the highest ionic conductivity[3]. Volume ratio of Ni:SDC was set for 60:40 to give the highest electrode performance. The core-shell type composite particles were prepared from an aqueous nitric acid solution containing cerium nitrate hexahydrate (Kojundo Chemical Laboratory, 99.9%), samarium nitrate hexahydrate (Mitsuwa's Pure Chemicals, 99.9%), and nickel acetate tetrahydrate (Kojundo Chemical Laboratory, 99.9%). The dispersed type composite particles were prepared from as aqueous nitric acid solution of almost the same as the core-shell type composite particles, while samarium oxide (Shin-Etsu Chemical, 99.9%) was used instead of samarium nitrate hexahydrate and ethylene glycol (Wako Pure Chemical Industries, 99.5%) was added to the solution. That aqueous solution for dispersed type composite particles was heat-treated, followed by dilution with water as a preprocessing. The details of the preprocessing procedure were described in the previous paper[8].

The spray pyrolysis apparatus consisted of an aqueous solution chamber with ultrasonic vibrators for mist generation, a carrier gas supplier (air was used as a carrier in this study), reaction furnaces, and a filter to capture the powders. Reaction furnaces were divided into four parts and their temperatures were set for 200, 400, 800, and 1000°C, respectively.

The synthesized powders were found to consist of spherical particles with about 1 μm in diameter from the SEM images, which were shown in the previous paper[5]. The core-shell type composite particles had hollows structure. Cerium was concentrated in the vicinity of the particle surface. Nickel was distributed uniformly inside the particles. The internal structure of the dispersed type composite particles was reported in detail in the previous paper[7]. The dispersed type composite particles were stuffed inside, and were composed of evenly distributed SDC and NiO having the size of around 10 nm.

Microstructure analysis

The lattice structures of the powders were measured by X-ray diffraction (XRD, XRD-6000, Shimadzu), and was confirmed to be composed of NiO and SDC. The microstructure variation from powders to cermet before fuel cell test was observed by SEM (S-2380N, Hitachi). The composite particles (Ni:SDC = 60:40 by volume) were sintered at the temperatures in the range between 1150 and 1350°C for 3 h after screen printing of the pastes with organic binder on lanthanum gallate based substrate $La_{0.8}Sr_{0.2}Ga_{0.8}Mg_{0.15}Co_{0.05}O_{2.8}$ (LSGMC, supplied as disk samples from Mitsubishi Materials). The morphology of the NiO-SDC and the distribution of Ce and Ni sintered at the temperatures in the range between 1150 and 1310°C was observed by TEM with energy-dispersive X-ray spectroscopy (EDS, KEVEX SIGMA LEVEL2) after Ar ion etching for obtaining a thin sample. TEM-EDS was selected instead of SEM-EDS to obtain the elemental distribution images with higher spatial resolution.

Single cell test

NiO-SDC/LSGMC/$Sm_{0.5}Sr_{0.5}CoO_{3-\delta}$ (SSC, Sakai Chemical Industry) cells (the electrode area: 2 cm^2) were fabricated and their electrochemical properties were measured at 750°C. NiO was reduced to be metallic Ni in H_2 atmosphere just before cell operation. SSC is one of the most attractive materials for cathode[9,10]. A glass ring was used as a sealing gasket at each compartment to avoid leakage of gases. The assembled cell was placed in the electric furnace. Dry hydrogen and dry air were supplied to the anode and cathode compartment, respectively, at sufficiently high flow rates of 200 ml/min. For the electrochemical characterization, the current-interruption method was used to separate the polarization of electrochemical reaction and IR drop from the total cell resistance.

RESULTS AND DISCUSSION

Microstructures of NiO-SDC sintered on substrate

The SEM images of the core-shell type NiO-SDC particles sintered on the electrolyte at various temperatures were shown in the previous paper[7]. The sintering at 1100°C made the particles start connecting to the neighbor particles, and the sintering at 1200°C brought about the formation of the network structures. The sample sintered at 1280°C showed wide channels and the shapes of the original particles disappeared.

The SEM images of the dispersed type NiO-SDC particles sintered on the electrolyte at various temperatures are shown in Fig. 1. The SEM image of the sample sintered at 1150°C indicated that the particles existed as a deposit on the electrolyte, and only small connections between the neighboring particles. The sample sintered at 1200°C began network-formation between the neighboring particles while the shape of the particles remained. Network-structure formation was progressed in the sample sintered at 1240°C, and the particles remaining the original shapes extremely decreased. When the sample was sintered at 1280°C, the original shapes in the particles disappeared. The grain size turned larger and the connection of the network turned wider after sintering at 1310°C.

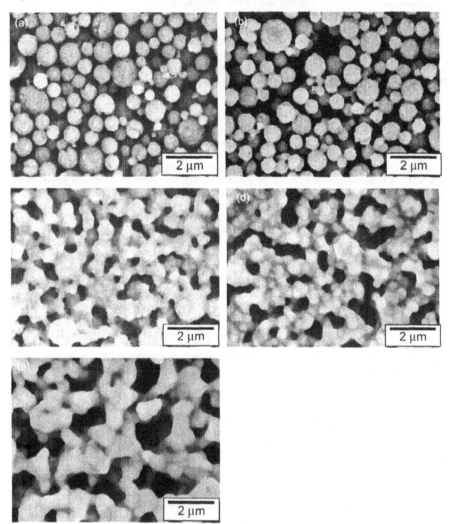

Figure 1. SEM images of the dispersed type NiO-SDC composite particles sintered at (a): 1150°C. (b): 1200°C. (c): 1240°C. (d): 1280°C. and (e): 1310°C for 3 h.

The distributions of NiO and SDC in the network are very important to understand the difference between core-shell and dispersed structures in the network-formation process on sintering of the composite particles, therefore, the microstructure and the distribution of NiO-SDC sintered at the various temperatures were analyzed by TEM-EDS.

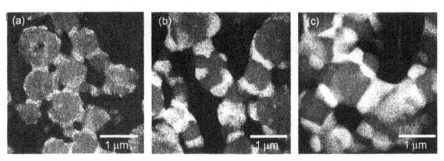

Figure 2. The elemental distribution images obtained from TEM-EDS analysis of NiO-SDC prepared from the core-shell structure particles sintered on the electrolyte at several temperatures for 3 h: (a) 1150°C. (b) 1200°C. and (c) 1280°C. The white and gray grains show SDC and NiO, respectively.

Fig. 2 shows the elemental distribution images of the samples prepared from the core-shell structure particles. The distributions of Ce, Sm and Ni were overlapped in the same image. The distributions of Ce and Sm showed the same patterns, and white signals mean the distributions of Ce and Sm. Gray signals mean Ni in NiO. The distribution of oxygen was confirmed to be oxides for Ce, Sm, and Ni. From the image of the sample sintered at 1150°C. it was found that the SDC grains on the surface of each particle connected at the early stages of sintering, that is, the aggregation of nickel can be controlled by the SDC grains between particles at the early stages of sintering. While the connections between the neighboring particles turned wider and the original shapes of the particles began to change after the sintering at 1200°C, the connections between the NiO grains were limited. The network structure of the NiO grains progressed during the sintering at 1280°C and the TPBs increased, which was good microstructure for the cell performance.

Fig. 3 shows the elemental distribution images of the samples prepared from the dispersed structure particles. The connections between the SDC grains and between the NiO grains in the neighboring particles were observed at 1150°C of the initial stage of the sintering, which was different from the behavior of the sample prepared from the core-shell structure particles. More connections between the grains were observed in the TEM-EDS image (Fig. 3a) than the SEM image (Fig. 1a). The SEM image shows the surface structure and the TEM-EDS image shows the internal structure, therefore, it was considered that more connections existed at inner part of the sample than at the surface. The original shapes of the particles remained in the TEM-EDS image as well as the SEM image. Fine primary grains existed and were more dispersed than those of the samples sintered at the same temperature from the core-shell structure particles. The original shapes of the particles disappeared by raising the sintering temperatures, however, the connecting necks forming network-structure did not become thicker than that prepared from the core-shell structure particles since SDC grains remained dispersed state. The connecting necks thickened by sintering at higher temperatures than 1300°C. It was found from these results that higher sintering temperature for the dispersed type particles was necessary to obtain the same network connecting necks than that for the core-shell type particles.

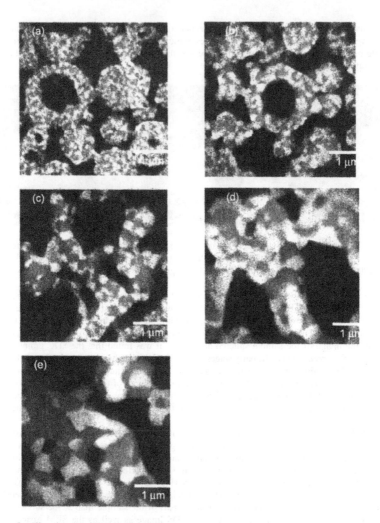

Figure 3. The elemental distribution images obtained from TEM-EDS analysis of NiO-SDC prepared from the dispersed structure particles sintered on the electrolyte at several temperatures for 3 h: (a) 1150°C. (b) 1200°C. (c) 1240°C. (d) 1280°C. and (e) 1310°C. The white and gray grains show SDC and NiO. respectively.

Cell performance variation with sintering temperatures

The performances of the cells with anodes sintered at the various temperatures were measured, and the variations of the IR drops and the anodic overpotentials with sintering temperatures were investigated by current interruption method. Fig. 4 shows the variations of the cell performance and the electrochemical properties with sintering temperature for the cells with anodes prepared from the core-shell type particles. The cathodic overpotentials were omitted in Fig. 4 as they were not related to the sintering temperature of the anodes and were confirmed to be constant. It was found that the IR drops decreased drastically with increasing the sintering temperatures, which was considered to occur increasing of connecting neck width, growing of network-structure formation, and the connecting between the NiO grains. The wider the connecting neck became and the more the network-structure grew, the shorter the average conducting paths turned. The IR drop is expected to decrease drastically due to the generation of Ni-Ni connection since Ni gives much higher electronic conductivity than that of SDC. The anodic overpotential decreased with the increasing of sintering temperatures. In general, apparent TPBs may tend to decrease with the increasing of sintering temperatures as the grain sizes of the components of the network increase. On the other hand, the effective TPBs may increase with sintering temperatures as electronic and ionic conduction are built by generating Ni-Ni and SDC-SDC connections. The result of the anodic overpotential variation indicated that the contribution of the latter case was superior for the anode prepared from the core-shell type particles. This is reasonable as Ni-Ni connections formed at higher temperature than 1200°C, though SDC-SDC connections formed even at 1150°C.

(a)   (b)

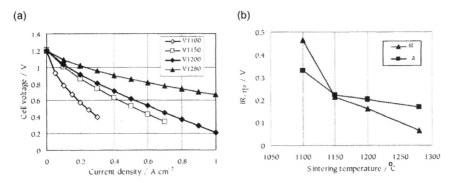

Figure 4. The cell performances at 750°C with anodes prepared from the core-shell type particles sintered at the various temperatures; (a) I-V curves, (b) IR drops and anodic overpotentials (IR free).

Fig. 5 shows the variations of the cell performance and the electrochemical properties with sintering temperature for the cells with anodes prepared from the dispersed type particles. The IR drops slightly decreased by increasing of sintering temperatures from 1200 to 1240°C, and were almost constant at the sintering temperatures higher than 1240°C. The reason why the IR drop decreased slightly at lower temperatures is that the IR drop sintered at 1200°C is not so large since some electron conducting paths exist by sintering at lower temperature due to Ni-Ni connections at the initial stages of the sintering in the case of the dispersed type composite particles (Fig. 3a). It was confirmed that the cell after sintering of the dispersed type particles at lower temperatures than 1240°C obtained lower IR drop than that prepared from the core-shell type particles. The possible cause of the almost constant IR drops in spite of higher temperature sintering is that the electron conducting paths do not increase so much since the dispersed SDC grains remain inside the connecting necks after sintering even at higher temperatures. The slight increase of the IR drop at 1310°C may be due to chemical reactions or ionic migration during high-temperature sintering[11]. The variation of the anodic overpotentials with sintering temperatures of anode prepared from the dispersed type composite particles showed that the anodic overpotentials slightly increased after sintering at the temperature higher than 1240°C, which was the opposite result of the cell with the core-shell type composite particles. This behavior suggested that the effect of decreasing apparent TPBs was remarkably visible since the rate of effective TPBs was relatively large due to the presence of Ni-Ni connections even at the initial stages of the sintering.

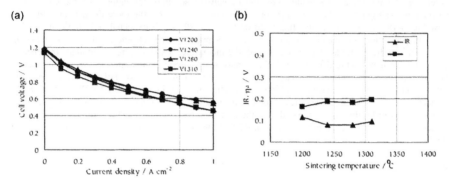

Figure 5. The cell performances at 750°C with anodes prepared from the dispersed type particles sintered at the various temperatures; (a) I-V curves. (b) IR drops and anodic overpotentials (IR free).

The difference in the network-structure formation processes with the internal structure of the particles

As was mentioned before. it was found that the network-structure formation processes from the composite particles were different depending on the internal structures of the particles. The different points of the processes between the core-shell type particles and the dispersed type particles were shown below and discussed.

(i) The connections of the particles at the initial stages of the sintering were different. In the case of the core-shell type particles. the SDC grains on the surface of the particles began to connect and the connection between the NiO grains started later. On the other hand. it was found that the connection between the NiO grains started at almost the same time as that between the SDC grains for the dispersed type particles.

(ii) The sintering temperatures demanded for the growth of the connecting necks are different. For the core-shell type particles. the network-structure began to be formed at around 1200°C. and the connections among the NiO grains and among the SDC grains were highly developed at 1280°C. For the dispersed type particles. the network-structure began to be formed at around 1240°C, and the connections among the NiO grains and among the SDC grains were sufficiently grown at the higher temperature than 1310°C.

(iii) The variations of electrochemical properties with sintering temperatures are different. The IR drops decreased drastically with the increasing of the sintering temperatures for the cells with the anodes prepared from the core-shell type particles, while the drastic decrease of the IR drop was not observed for the cells with the anodes prepared from the dispersed type particles. In the case of the core-shell type particles, only the SDC grains on the surface of the composite particles could be connected at the initial stages of the sintering, and the connection of the NiO grains increased with increasing of the sintering temperatures, therefore the IR drops decreased drastically with increasing of the sintering temperatures. While the connections among the NiO grains were observed at the initial stages of the sintering for the dispersed type particles. therefore. the IR drop was not so large after sintering even at the lower temperature of 1200°C. The anodic overpotentials decreased with increasing of the sintering temperatures for the core-shell type particles, which indicated that the contribution of the increasing of the effective TPBs was larger than that of the decreasing of the apparent TPBs. The anodic overpotentials for the dispersed type particles increased slightly with the increasing of sintering temperatures. This behavior suggested that the effect of decreasing apparent TPBs was large since the range of effective TPBs was relatively high even at the initial stages of the sintering.

When the core-shell type composite particles are used, the connections among SDC grains generate and the NiO grains inside the particles grow at the initial stages of the sintering, and the densification of the anode can be controlled. Therefore, the network-structure having wider connecting necks can be formed after high temperature sintering, and high-performance anode can be obtained. It can be said that the use of the core-shell type composite particles is effective for the usage demanding relatively high heat-treatment temperatures for example the co-sintering with the electrolyte as anode-supported cells. When the dispersed type composite particles are used, electronic conduction paths have existed at the initial stages of the sintering since the connections among the NiO grains generate on the surface of the particles. Furthermore, large TPBs are obtained due to the well-dispersion of the SDC and the NiO grains, therefore, the particles have an advantage that high-performance can be obtained by sintering at lower temperatures. It can be said that the use of the dispersed type composite particles is effective for the usage demanding the sintering at the temperature as low as possible.

Lastly, it is indicated that the applying of spray pyrolysis method on the synthesis of the composite particles for SOFC anode material is effective since it can individually realize the composite particles having the microstructure optimized for each application.

CONCLUSIONS

The NiO-SDC composite powders having the core-shell type structure and the dispersed type structure were synthesized by spray pyrolysis method. The network-structure formation process of the NiO-SDC composite particles was analyzed in detail. It was found that the network-structure formation processes from the composite particles were different depending on the internal structures of the particles. In the case of the core-shell type particles, the SDC grains on the surface of the particles began to connect and the connection between the NiO grains started later. On the other hand, it was found that the connection between the NiO grains started at almost the same time as that between the SDC grains for the dispersed type particles. For the core-shell type particles, the network-structure began to be formed at around 1200°C, and the connections among the NiO grains and among the SDC grains were highly developed at 1280°C. For the dispersed type particles, the network-structure began to be formed at around 1240°C, and the connections among the NiO grains and among the SDC grains were sufficiently grown at the higher temperature than 1310°C. The IR drops decreased drastically with increasing of the sintering temperatures for the cells with the anodes prepared from the core-shell type particles, while the drastic decrease of the IR drop was not observed for the cells with the anodes prepared from the dispersed type particles. When the core-shell type composite particles are used, the connections among the SDC grains generate and the NiO grains inside the particles grow at the initial stages of the sintering, and the densification of the anode can be controlled. Therefore, the network-structure having wider connecting necks can be formed after high temperature sintering, and high-performance anode can be obtained. When the dispersed type composite particles are used, electronic conduction paths have existed at the initial stages of the sintering since the connections among the NiO grains generate on the surface of the particles. Furthermore, large TPBs are obtained due to the well-dispersion of the SDC and the NiO grains, therefore, the particles have an advantage that high-performance can be obtained by sintering at lower temperatures.

REFERENCES

[1] T. Ishihara, H. Matsuda, and Y. Takita, Doped LaGaO Perovskite Type Oxide as a New Oxide Ionic Conductor, *J. Am. Chem. Soc.*, **116**, 3801-3 (1994).

[2] K. Huang, M. Feng, and J. B. Goodenough, Sol-Gel Synthesis of a New Oxide-Ion Conductor Sr- and Mg-Doped LaGaO$_3$ Perovskite, *J. Am. Ceram. Soc.* **79**, 1100-4 (1996).

[3] H. Inaba and H. Tagawa, Ceria-Based Solid Electrolytes, *Solid State Ionics* **83**, 1-16 (1996).

[4] R. Maric, S. Ohara, T. Fukui, H. Yoshida, M. Nishimura, T. Inagaki, and K. Miura, Solid Oxide Fuel Cells with Doped Lanthanum Gallate Electrolyte and LaSrCoO$_3$ Cathode, and Ni-Samaria-Doped Ceria Cermet Anode, *J. Electrochem. Soc.* **146**, 2006-10 (1999).

[5] S. Ohara, R. Maric, X Zhang, K. Mukai, T. Fukui, H. Yoshida, T. Inagaki, and K. Miura, High performance electrodes for reduced temperature solid oxide fuel cells with doped lanthanum gallate electrolyte: I. Ni–SDC cermet anode, *J. Power Sources*, **86**, 455-8 (2000).

[6] H. Yoshida, T. Inagaki, K. Hashino, M. Kawano, H. Ijichi, S. Takahashi, S. Suda, and K. Kawahara, Study on the Network-Formation Process in the Sintering of the NiO-SDC Composite Powders Prepared by Spray Pyrolysis Method, *in Solid Oxide Fuel Cells IX*, S.C. Singhal and J. Mizusaki, Editors, **PV 2005-07**, The Electrochemical Society Proceedings Series, Pennington, PJ, 1382-89 (2005).

[7] H. Yoshida, H. Deguchi, M. Kawano, K. Hashino, T. Inagaki, H. Ijichi, M. Horiuchi, K. Kawahara, and S. Suda, Study on pyrolysing behavior of NiO-SDC composite particles prepared by spray pyrolysis technique, *Solid State Ionics* **178**, 399-405 (2007).

[8] S. Suda, K. Kawahara, M. Kawano, H. Yoshida, and T. Inagaki, Preparation of Matrix-Type Nickel Oxide/Samarium-Doped Ceria Composite Particles by Spray Pyrolysis, *J. Am. Ceram. Soc.*, **90**, 1094-1100 (2007).

[9] T. Ishihara, M. Honda, T. Shibayama, H. Minami, H. Nishiguchi, and Y. Takita, Intermediate Temperature Solid Oxide Fuel Cells Using a New LaGaO$_3$ Based Oxide Ion Conductor, *J. Electrochem. Soc.*, **145**, 3177-83 (1998).

[10] K. Kuroda, I. Hashimoto, K. Adachi, J. Akikusa, Y. Tamou, N. Komada, T. Ishihara, and Y. Takita, Characterization of solid oxide fuel cell using doped lanthanum gallate, *Solid State Ionics*, **132**, 199-208 (2000).

[11] J. Yan, H. Matsumoto, M. Enoki, and T. Ishihara, High-Power SOFC Using La$_{0.9}$Sr$_{0.1}$Ga$_{0.8}$Mg$_{0.2}$O$_{3-\delta}$/Ce$_{0.8}$Sm$_{0.2}$O$_{2-\delta}$ Composite Film, *Electrochem. Solid-State Lett.*, **8**, A389-91 (2005).

FUNCTIONALLY GRADED COMPOSITE ELECTRODES FOR ADVANCED ANODE-SUPPORTED, INTERMEDIATE-TEMPERATURE SOFC

Juan L. Sepulveda, Raouf O. Loutfy, and Sekyung Chang
Materials and Electrochemical Research (MER) Corporation
Tucson, Arizona 85706, USA
and
Peiwen Li, and Ananth Kotwal
Department of Aerospace and Mechanical Engineering
The University of Arizona
Tucson, AZ 85721

ABSTRACT

This paper describes the development and modeling of anode-supported intermediate-temperature solid oxide fuel cells (ACN-AS-IT-SOFC) that exhibit high electrochemical efficiency, high degree of fuel utilization, and low operating temperature characteristics. The proposed cell design is fuelled by hydrogen or in-situ reformed fuel and operates at a lower temperature of 600-800°C producing a maximum power density of 2-2.2 W/cm^2. The innovative design for the ACN-AS-IT-SOFC fuel cell makes use of a porous anode consisting of a combination of a highly conductive anode capillary network (ACN) running through the supporting anode manufactured using MER poly capillary material technology. The highly porous anode allows for free fuel gas access to the functional anode. Operating at low temperature of 600-800°C it allows the use of less expensive interconnect materials such as ferritic steels. A method to identify over-potentials caused by different polarizations in an SOFC with multi-layer hybrid electrodes is also presented. The contributions of each polarization to the total loss in a fuel cell can be identified. The polarization causing the maximum over-potential is then considered as the primary source of internal losses, and optimization is focused to improve the power density. The analysis for the mass transfer polarization considers bulk convection and diffusion in porous layers from bulk flow to the interface of the electrode and electrolyte. Values of the exchange current densities are determined empirically by matching analytical and experimental results. Effects of porosities and thicknesses of anode, cathode, and functional graded layers are modeled and optimized to attain maximum power density.

INTRODUCTION

Current SOFC development aims to design cells able to operate at low temperature and under high levels of electrochemical efficiency. MER's approach to achieve these objectives is to design and develop anode-supported intermediate-temperature solid oxide fuel cell (ACN-AS-IT-SOFC) that exhibits high electrochemical efficiency, high degree of fuel utilization, and low operating temperature characteristics intended for defense applications or other commercial users. The SOFC cell design under development is fuelled by hydrogen or by in-situ reformed fuel like JP-8 or diesel. It will operate at a lower temperature of 600-800°C (as compared to 800-1000°C for conventional SOFC cells). It is anticipated to produce a maximum power density of 2.5 W/cm^2 when operating at 0.55 V, 4.60 amp/cm^2.

The innovative design for the MER's ACN-AS-IT-SOFC fuel cell makes use of a highly electrically conductive porous anode consisting of a combination of a highly conductive anode capillary network (ACN) embedded in a more conventional porous Ni/electrolyte supporting a thin functional anode layer consisting of Ni/electrolyte, a thin electrolyte layer (scandia stabilized zirconia ($Zr_{0.9}Sc_{0.1}O_{1.95-x}$) ($S_{10}SZ$) or scandia ceria stabilized zirconia ($Zr_{0.89}Sc_{0.10}Ce_{0.01}O_{1.95-x}$) ($S_{10}C_1SZ$)), coated with a functional cathode layer consisting of LSCF/electrolyte, and a current collector layer consisting of LSCF. The capillary network running through the supporting anode is manufactured using MER's poly-capillary-material technology (PCM) and is made of the same material composition used for functional anode enhancing H diffusion and providing direct contact in between the thin functional anode and the anode interconnect. The balance of the anode consists of porous Ni/electrolyte cermet allowing free fuel gas access to the functional anode. Fuelled by pure hydrogen or hydrogen containing reformed JP-8 or diesel fuel, this cell configuration is designed to operate at low temperature of 600-800°C which allows the use of less expensive interconnect materials such as ferritic steels, or Ni alloys, or silver alloys. $S_{10}SZ$ or $S_{10}C_1SZ$ are used as electrolyte as specified in the stack cell composition described in Table I. Experimentally these stacks are built and tested using button cell technology. The schematic of the configuration of the unit cell electrode is depicted in Figure 1.

Table I. Anode Supported Unit Cell Composition.

	Cell Configuration 1	Cell Configuration 2
Cathode Current Collector Layer(CCCL)	LSCF (50-60 μm)	LSCF (50-60 μm)
Cathode Functional Layer (CFL)	LSCF/S_{10}SZ (20-30 μm)	LSCF/$S_{10}C_1$SZ (20-30 μm)
Electrolyte Layer (EL)	S_{10}SZ (7-10 μm)	$S_{10}C_1$SZ (7-10 μm)
Anode Functional Layer (AFL)	Ni/S_{10}SZ (10-20 μm)	Ni/$S_{10}C_1$SZ (10-20 μm)
Anode Support (AS)	Ni/S_{10}SZ (500-1000 μm)	Ni/$S_{10}C_1$SZ (500-1000 μm)
Anode Capillary Network (ACN)	Ni/S_{10}SZ(1000 μm, 60μmΦ)	Ni/$S_{10}C_1$SZ(1000μm, 60 μm Φ)

S_{10}SZ: Scandia Stabilized Zirconia: $Zr_{0.9}Sc_{0.1}O_{1.95-x}$
$S_{10}C_1$SZ: Scandia and Ceria Stabilized Zirconia: $Zr_{0.89}Sc_{0.16}Ce_{0.01}O_{1.95-x}$
LSCF: Lanthanum Strontium Cobalt Ferrite: $La_{0.6}Sr_{0.4}Co_{0.2}Fe_{0.8}O_{3-\delta}$
CFL: consists of 50/50% by weight LSCF/Electrolyte
AFL: consists of 50/50% by weight NiO/Electrolyte plus 0-10% Carbon
AS: consists of 70/30% by weight NiO/Electrolyte plus 5-25% Carbon, 30%-40% porosity
ACN: consists of 70/30% by weight NiO/Electrolyte 45% porosity

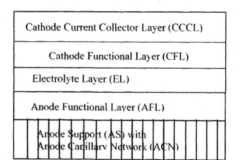

Figure 1. MER's Anode Supported SOFC Structure

Anode-supported intermediate-temperature solid oxide fuel cells are being pursued by MER for production at plant-scale, since they present several advantages over conventional SOFCs [1-3]. They can be operated at lower temperature (800°C) allowing the use of lower cost materials for interconnect and manifold, reducing sealing and corrosion problems, and increasing lifetime and reliability. However, operating at lower temperatures requires an increase in the electrochemical performance. This is accomplished by using advanced electrolyte and cathode compositions, deposited as very thin layers, and by optimizing the microstructure of the cell layers. This way, the electrical resistance of the electrolyte is minimized. The net result is enhanced cell performance that has been reported to produce a maximum power density of 1.8 W/cm^2 (0.55 V at 3.34 A/cm^2) when using a standard formulation consisting of Ni for the conventional porous anode support and functional anode, Y_8SZ as the electrolyte and LSM for the functional cathode and cathode current collector after optimizing the porosity and the thickness of the different layers[1]. Substantially improved performance is expected to be obtained from the fuel cell formulation under development at MER since a highly conductive anode capillary network is inserted in the porous anode, a higher ionic conductivity electrolyte, and a higher catalytic activity cathode are also used. As shown in Table II, the electrolytes used for this SOFC design exhibit 4-7 times higher conductivity than Y_8SZ at 750°C operating temperature. Improvements provided by the LSCF cathode will complement the ACN-AS-IT-SOFC performance[4, 5]. A maximum power density of 2.5 W/cm^2 (0.55 V at 4.60 A/cm^2) is expected for the cell proposed in this project although this estimate is very conservative. This ACN-AS-IT-SOFC will be developed for reformed H containing diesel, JP-8, or other hydrocarbon fuels. The fuel and oxidants are manifolded and can be pressurized to improve the performance of the cell.

Table II. Electrical Conductivity and CTE of Electrolyte Materials for SOFC. σ_i (S/cm)

	$G_{10}DC^6$	$S_{15}DC^6$	$S_{10}C_1SZ^8$	$S_{10}SZ^7$	Y_8SZ^6
1000°C	0.380	0.263	0.310	0.270	0.100
800°C	0.126	0.100	0.130	0.090	0.030
750°C	0.100	0.079	0.088	0.052	0.013
700°C	0.063	0.050	0.058	0.020	0.006
650°C	0.039	0.031	0.030	0.012	0.004
600°C	0.031	0.025	0.019	0.006	na
CTE 25-1000°C (ppm/°K)	13.4	12.7	12.1	12.0	10.7

Table III. Typical Physical Properties of ACN AS-IT-SOFC Unit-Cell Components[2-8]

Component	Conductivity at 800°C (S/cm)	Thickness, (μm)	Porosity (%)	CTE, (ppm/°K)
CCCL: LSCF	380	12	35	14.2
CCCL: LSC	1100	12	32	18.8
CCCL: LSM	152	13	36	13.0
CFL: LSM-YSZ	93	25	40	11.5
CFL:LSCF-$G_{10}DC$	120	25	45	14.0
CFL:LSCF- $S_{10}C_1SZ$	130	20-30	45	14.1
CFL:LSCF- $S_{10}SZ$	130	20-30	45	14.1
EL: $S_{10}C_1SZ$	0.130	8	<3	12.1
EL: $S_{10}SZ$	0.090	8	<3	12.0
EL: $G_{10}DC$	0.126	8	<3	13.4
EL: Y_8SZ	0.030	6-8	<3	10.5
AFL: Ni-Y_8SZ	1050	15-20	9.7	14.0
AFL: Ni- $S_{10}C_1SZ$	1100	20	10	14.0
AFL: Ni- $S_{10}SZ$	1100	20	10	14.0
AS: Ni- Y_8SZ	1000	1000	42	14.0
AS: Ni- $S_{10}C_1SZ$	1090	1000	55	14.0
AS: Ni- $S_{10}SZ$	1100	1000	57	14.0
ACN: Ni- $S_{10}C_1SZ$	1400	25-50 Φ	0	14.0
ACN: Ni- $S_{10}SZ$	1450	25-50 Φ	0	14.0

Table II summarizes the electrical conductivity and coefficient of thermal expansion for high conductivity electrolytes including the electrolytes used in the MER SOFC stack. All these ceramics exhibit high ionic conductivity. Similarly, Table III summarizes typical physical properties of ACN-AS-IT-SOFC unit-cell components including those used by MER. By analyzing the data provided in Table II and Table III, it can be concluded that several alternative design are possible to manufacture high performance ACN-AS-IT-SOFC although MER is focusing on the designs shown in Table I exclusively.

For the ACN AS-IT-SOFC to operate efficiently it is necessary to minimize the total polarization losses, including electrode and ohmic losses[1]. The total loss will depend on the ionic conductivity of the electrolyte, the electronic and ionic conductivities of the two electrodes, the thicknesses of the electrolyte, the thicknesses of the electrodes, and ohmic resistances of the interfaces. For ACN-AS-IT-SOFC the ohmic loss associated to electrolyte thickness is minimized. A thin, high ionic conductivity electrolyte is preferred for the ACN-AS-IT-SOFC design to minimize ohmic losses. Y_8SZ electrolyte has been used extensively due to its excellent stability in both reducing and oxidizing environments. However, the electrical conductivity of Y_8SZ is lower than other materials recently developed such as $G_{10}DC$, LSGM, $S_{10}SZ$, and $S_{10}C_1SZ$. Good potential has been shown for $S_{10}SZ$, and $S_{10}C_1SZ$[9]. The use of $G_{10}DC$ requires protection of the reduction environment on the anodic side of the cell which could be accomplished by layering with $S_{10}SZ$[9]. For the current MER design the use of $S_{10}SZ$ or $S_{10}C_1SZ$ is preferred although several other high ionic conductivity electrolytes are possible.

$S_{10}C_1SZ$, a highly conductive novel electrolyte for SOFC at temperature of 700-800°C, consists of ZrO_2 co-doped with 10 m/o Sc_2O_3 and 1 m/o CeO_2[8]. This electrolyte does not show any phase transition during heat treatment up to 1550°C, is stable as a cubic phase in all temperature ranges, and shows much higher electrical conductivity than Y_8SZ in the 600-1000°C temperature range as shown in Table II.

The ohmic loss will also depend on the conductivity of the electrodes and sheet resistance. Therefore higher conductivity and lower sheet resistance are desirable. This is why an ACN for the anode is included in the MER design. The ohmic loss will also depend on the relative amounts of the two phases present, the porosity, and its microstructure. Electrode porosity, pore size distribution, and the pore morphology will affect gas transport through the electrodes and concentration polarization. Also very importantly, the CFL and AFL morphology, including the three phase boundary (TPB) length, will have an effect on the activation polarization. The total length of the TPB has to be as large as possible. The electrode thickness will also affect the concentration polarization. The thicker the support electrode, the greater the concentration polarization. Therefore, MER uses a capillary anode network to minimize this effect and improve electrochemical cell performance. However, sufficient thickness is needed to provide appropriate strength and mechanical integrity, but for thinner electrodes the concentration polarization is lower. Target porosity and thickness for the different layers used are indicated in Table I and Table III. Porosity and thickness could be slightly varied to optimize performance.

In previous research, this group of researchers[9] has found best fuel cell performance was achieved when using the $S_{10}SZ$ and $S_{10}C_1SZ$ electrolytes in combination with NiO-electrolyte 70%/30% by weight mix for the anode and LSFC-electrolyte 50%/50% by weight mix for the cathode. LSFC has been found most suitable to operate at 600-800°C because of its high mixed ionic and electronic conductivity and its excellent thermal and chemical compatibility[4]. LSCF has also been successfully used with $G_{10}DC$ electrolyte and as: LSCF-$G_{10}DC$ (40 vol%), LSCF-Pt (0.5-1.0 vol %), or LSCF-$G_{10}DC$ (40 vol%)-Pt (1 vol%) in cathode functional layer (CFL).

The MER SOFC focuses on the implementation of ACN assisted AS-IT-SOFC technology for plant-scale production. The use of PCM technology to manufacture the anode support presents several performance and manufacturing advantages to scale up fuel cell production. Green forming and cutting the porous anode components to a given shape, coupled with a large capacity spray coater facilitates large volume production of ACN-AS-IT-SOFC at a reduced cost. Cutouts and perforations, as required by cell geometry can be imparted on the electrolyte in the green state using a laser cutter unit.

Anode and cathode coatings are accomplished using a spin coater. Materials to be coated are suspended in a vehicle to generate an ink. This technique is particularly suited for the metallization of flat SOFC components and rapid prototyping of electrodes. Alternatively, screen printing could also be used.

Interconnects made out of Pt or Ag are screen printed on top of the anode and cathode as interconnect. Ferritic steel could also be used for interconnect, especially when scaling up to larger size stacks. The ACN-AS-IT-SOFC technology is feasible at the proposed operating temperature of 800°C. At this temperature, the metal components used consisting of stainless steel, ferritic steel, Ni, or Ni-based superalloys, will have good longevity. The use of high Cr containing iron alloy will not be needed given that the temperature is kept at 800°C. Croffer 22 alloy could also be used for increased cell longevity.

MODELING

To understand the effect of the functionally graded electrode thicknesses and porosities on the fuel cell performance, the over-potentials due to mass transfer resistance, ohmic loss, and activation polarization need to be identified to be able to draw conclusions. The analytical electrochemical model developed has been explained in detail previously[10, 11]. The concentration polarization is virtually reflected in the second term in the Nernst equation for the electromotive force. Because the analysis for optimization of fuel cell components involves processes including mass transfer, current conduction and activation polarization, the complexity of modeling must be manageable in order to focus on the important factors. For this consideration, a zero-dimensional mass transfer model is proposed to identify the concentration polarization. Consider the mass transfer of oxygen on the cathode side for example. For a SOFC with graded functional layers, the mass transfer of oxygen from bulk flow to the electrode/electrolyte interface experiences three processes, convective mass transfer, diffusion through the porous anode layer, and diffusion through the anode functional layer. Similar analytical procedures are applicable to hydrogen and water vapor in the anodic side. The electrical circuit network is applied to analyze the ohmic resistances and potential loss during the current collection and the Butler-Volmer equation is considered to calculate the activation polarization.

The model was verified using a button cell experimental data set reported by Jung et al.[3] Exchange current densities of 12000 A/m^2 and 24000 A/m^2 for the cathode and the anode respectively were used. Experimental and simulated polarization and power density curves were in good agreement validating the applicability of the model as shown in the corresponding example of Figure 2.

This model was used to identify over-potentials caused by different polarizations in an SOFC with functional graded composite electrodes[10]. The contributions of each polarization to the

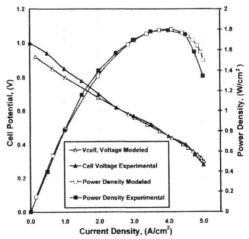

Figure 2. Model Verification using Polarization Curves, V-I Curves[3].

total loss in a fuel cell can be identified and compared. Special attention can be given to the polarization causing the maximum over-potential and then measures can be taken to reduce the internal losses to a minimum this way optimizing the power density. The analysis for the mass transfer polarization fully considers the mass transfer processes, including bulk convection, and diffusion in porous layers from bulk flow to the electrode/electrolyte interface. A current conduction circuit is used to consider the ohmic losses. For the activation polarization, values of the exchange current densities are determined empirically by fitting the simulated analytical results to the experimental results.

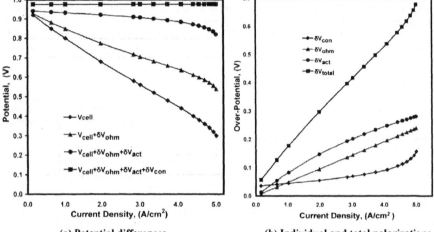

(a) Potential differences   (b) Individual and total polarizations
Figure 3. Potential Losses due to the Different Polarizations.

For the same SOFC discussed above, the three types of polarizations are shown in Figure 3. The concentration polarization shows sharp increase, when the current density is above 4.0 A/cm². The ohmic polarization increases linearly with the increase of current density. The concentration polarization is found to be lower than that of ohmic polarization. In a SOFC, the operating temperature is typically in between 600°C and 1000°C and thus the molecular gas diffusion is relatively strong, which results in less concentration polarization. Although the exchange current densities are dramatically improved due to the adoption of the graded functional layers, the activation polarization still takes the largest portion of the total losses in the SOFC. This is a very important indication that the effort to reduce activation polarization in SOFCs is still of significance to the total reduction of the losses.

In this study, functionally graded electrode combinations are being developed and were used to determine the effects of porosity and thickness for the anode, cathode, and functional electrode layers using model simulations. At the same time, the power density can be optimized. The composite electrode consists of layers in which the electrodes and the electrolyte mix and overlap providing a sizable triple-phase-boundary (TPB) length increase. Using functional electrode layers also contributes to several times increase of the exchange current density, reducing the activation polarization dramatically. Haanappel et al.[2] have reported a power density of 1.05 W/cm² at 800°C for a SOFC with cathode interlayer by LSM and YSZ at 50 wt.% ratio. Zhao and Virkar[1] fabricated a button-sized SOFC with YSZ electrolyte, Ni–YSZ anode support, Ni–YSZ anode functional layer, LSM–YSZ cathode functional interlayer, and LSM current collector. They reported a maximum power density of

1.8W/cm^2 at 800°C. These approaches are similar to the one being pursued by MER shown in Table I. In order to increase the diffusivity of the thick anode layer, capillary network is fabricated on anode layer. A number of studies using graded multi-layer SOFC designs have also been reported in the literature [3, 12-14].

The best balance in between thicknesses and porosities of the functional layers and electrode layers has been an issue of interest to SOFC developers[1, 2]. For the anode and cathode layers, a larger thickness may be helpful to support the fuel cell and also helpful to reduce the ohmic loss, since current must conducts in the in-plane or lateral direction in the electrodes. However, thicker electrode layers may incur larger mass diffusion resistance and is thus not desirable from the mass transfer point of view. It is necessary then to determine the best thickness for the electrode layers and hence to reduce the internal losses. The porosity of the electrode layers also has to be optimized. From the mass transfer point of view, a high porosity in electrodes is favorable for species to permeate through. However, the current conduction capability of the electrodes decreases with increasing porosity. Therefore, optimizing the porosity for electrodes in a SOFC will be important when trying to optimize the power density.

Parameters for the analytical simulation corresponding to conditions of interest for this study are summarized in Table IV and Table V. A contact resistance of 5.0×10^{-3} Ωcm^2 between a current collector and an electrode was used in the analysis. Hydrogen fuel at a stoichiometric coefficient of 1.2 and air at the oxygen stoichiometric coefficient of 2.0 are considered at the fuel cell operating temperature of 800°C.

Figure 4 shows the power densities for different anode thicknesses for four different anode porosities. At low porosity, thickness variation causes significant difference in the power density. Evidently, reducing the anode thickness results in greater power density. However, this trend becomes insignificant when the anode porosity increases. At 20% anode porosity, a reduction in anode thickness from 1100μm to 200μm results in 12.8% increase in power density. However, at 50% anode porosity, only 2.8% power density increase is observed. These analytical results are of great significance. Depending on the electrode porosity selected during fabrication, the proper anode thicknesses may be selected, which may be thicker for increased strength, allowing the used of a higher porosity. Conversely, the electrode could be thinner, if lower porosity is used. In the experimental study at MER, a highly electrically conductive porous anode has been developed, which is a combination of a highly conductive anode with capillary network (ACN), which helps to maintain high porosity in the anode

Table IV. Components Thickness and Porosity.

Cathode Current Collector Layer (CCCL)		Porosity
LSCF	60 μm	0.32
Cathode Functional Layer (CFL)		
LSCF/S$_{10}$SZ	30 μm	0.40
Electrolyte Layer (EL)		
S$_{10}$SZ	10 μm	< 0.03
Anode Functional Layer (AFL)		
Ni/S$_{10}$SZ	20 μm	0.10
Anode Support (AS)		
Ni/S$_{10}$SZ	1000 μm	0.55

Table V. Material Conductivity.

Component/Material	Conductivity at 800°C σ (S/cm)
(CCCL) LSCF	380
(CFL) LSCF/S$_{10}$SZ	130
(EL) S$_{10}$SZ	0.090
(AFL) Ni/S$_{10}$SZ	1100
(AS) Ni/S$_{10}$SZ	1400

(For porous media[15, 16], conductivity $\sigma_{eff} = \sigma(1-\varepsilon)^{1.5}$

support layer. Because of the improved porosity, a thicker anode is used for better strength of the SOFC.

Anode support green structures consisted of a mixture of NiO and electrolyte following the formulation shown in Table I. PVA 8% by volume was used as binder for the green structure and carbon (lamp black) was used at 0-6 wt.% as pore generator to increased porosity. 2 1/8" outside diameter (OD) anode support were produced after dry pressing in a Carver press. The anode supports were spin coated with a FA and EL layers, which compositions are described in Table I, and sintered in air for two hours at 1400°C. Fired density reached 70-75% of theoretical indicating that 30-25% porosity was typically attained. The shrinkage during sintering was 18-20%. After sintering, the electrolyte was spin coated with the CFL and CCCL and then fired again for two hours at 1200°C.

The simulation results showed in Figure 5 shows the effect of anode porosity on power density. These results confirm the hypotheses proposed by the MER design. Simulated results indicate an increase in porosity from 20% to 35% will result in a 10% increment of maximum power density.

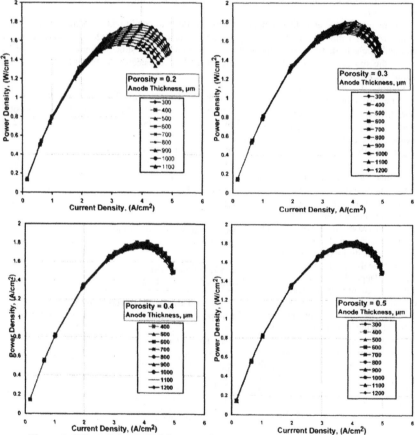

Figure 4. Effect of Anode Thickness on Power Density at Four Different Porosities.

When the porosity is above 40%, any effort to increase the porosity will not make a significant power density increase. When porosity is high, 40% to 50%, the change of anode thickness has insignificant effect on the power density increase. If the porosity is low, about 20%, changes of anode thickness can have significant effect on the power density. Although the improvement of anode porosity is helpful to the mass transfer and hence to the reduction of concentration polarization, it may not be similarly effective when the porosity is higher than 40%.

EXPERIMENTAL

The electrochemical performance for the cells was tested using a button cell experimental setup. The button cell experimental setup was provided with a pneumatic control system to feed hydrated H and air. Hydrogen is percolated through a water column to provide approximately 3% moisture content. By controlling the temperature of the water and the length of the column it is possible to change the water content. The button cells were built using zirconia tubes fitted with the electrode at one end using high temperature ceramic cement. A zirconia tube of 1.75" in diameter was used for all experiments.

Porous anode supports using three different types of electrolyte were built. Highly porous and micro-capillary network anodes were produced and coated with the corresponding AFL, EL, CFL, and

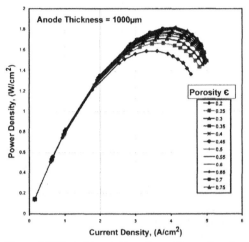

Figure 5. Effect of Anode Porosity on Power Density at a Constant Anode Thickness.

CCCL. Both AS and CCCL external faces were coated with Pt paste and fired. Finally, Pt wires were bonded to both AS and CCCL faces to function as interconnect to connect to the external circuit. Fully assembled button cells were prepared and operated at temperatures as high as 800°C to get corresponding polarization curves. A computer assisted data acquisition system was used in the experimental setup.

All PCM structures were built using paraffin wax. Microcapillary network of 50-70 μm in diameter were obtained. At the same time, porous anode fabrication was carried out in-tandem using uniaxial dry pressing of powder precursor components to generate the green (unfired) anode support. Anode support green structures consisted of a mixture of NiO and electrolyte following the formulation shown in Table I. PVA 8% by volume was used as binder for the green structure and carbon (lamp black) was used at 0-6 wt.% as pore generator to increased porosity. 2 1/8" outside diameter (OD) anode support were produced after dry pressing in a Carver press. The anode supports were spin coated with a FA and EL layers, which compositions are described in Table 1, and sintered in air for two hours at 1400°C. Fired density reached 70-75% of theoretical indicating that 30-25% porosity was typically attained. The shrinkage during sintering was 18-20%. After sintering, the electrolyte was spin coated with the CFL and CCCL and then fired again for two hours at 1200°C.

Ink formulations for AFL, EL, CFL, and CCCL were prepared using the same experimental procedure consisting of ball milling in with high purity ZrO_2 grinding media using terpineol and ethyl cellulose as the vehicle. Figure 6 shows some representative examples of the electrodes produced prior to being built into button cells. The fired anode ceramic microstructure of Figure 6 shows the ACN microcapillary details. The capillary network was 50-70 μm in diameter.

Figure 7 and Figure 8 shows cross sections of the electrodes produced as depicted by SEM and elemental EDS mapping. Thickness of the different layers was evaluated after performing EDS mapping for the different elements contained in the precursor powders. A dense electrolyte layer approximately 7 μm thick was typically obtained. The amount of carbon can be varied to control porosity adjacent to the EL layer.

Heraeus Cermalloy Pt paste CL-11-5100 was applied to the external face of the AS and CCCL layers for current collection purposes and was bonded to Pt wires used as interconnect to the external circuit. Two Pt wires were bonded to the AS and two Pt wires were bonded to the CCCL to determine voltage and to connect to the external circuit. 1 mm OD Pt wires were used for this last task. Figure 9 shows some of the finished electrode with wires in place. and bonded to the tube. Figure 10 shows an example of a finished cell. Figure 11 shows a cell during testing at temperature.

Several electrodes have been successfully built with all four additional layers in place as desired. Very good electrical conductivity was observed across the CCCL under ambient temp conditions when using an ohm meter. The Pt metallization and Pt wire bonding were also accomplished satisfactorily. Strong metallization bonding was obtained with very good electrical

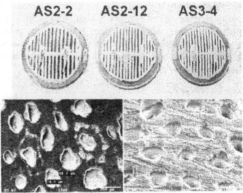

Figure 6. Anode supported electrodes built via dry pressing and PCM microcapillarity.

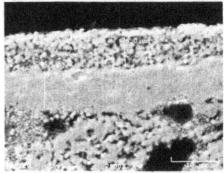

Figure 7. AS1-2 electrode microstructure.

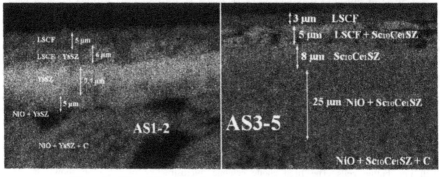

Figure 8. AS multilayer electrode cross-section EDS mapping.

conductivity. The Pt wire bonding to both anode and cathode were very strong and satisfactory. Wrap around Pt metallization was used to connect the anode to the front lead interconnect, this way, simplifying connections to the external circuit. Examples are provided in Figure 9 and Figure 10. Assembly of the button cells was also very successful. In excess of ten cells have been fully assembled so far. Bonding of the electrodes to the zirconia tube was accomplished using high temperature ceramic cement.

EXPERIMENTAL RESULTS

All the cells exhibited a working life of more than two hours at temperature (800°C). However, the electrical performance of the experimental cell runs has been less than desirable and has not allowed for the experimental evaluation of the cell performance.

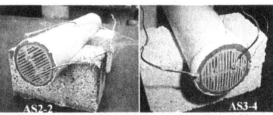

Figure 9. Electrodes bonded ready for cell assembly.

Experimenting with the assembled cells indicated that an acceptable electrical isolation in between the cathode and the anode has not been fully accomplished yet. Although under ambient conditions very high resistance across the electrodes (or open circuit conditions) was obtained, at experimental running temperature conditions during testing, it became apparent that the anode and the cathode were not fully isolated. The resistance across the electrode went from 5 MΩ to 0 Ω when the temperature increased from ambient to 800°C. Under these conditions, and as

Figure 10. AS3-4 assembled cell ready for testing.

expected, no voltage was detected for all this first generation of cells except a low voltage (0.2-0.3 V) at low temperature conditions (150°-200°C). It is apparent that the cathode and the anode were short circuited across the electrolyte. In some of the electrodes, a very small electrical conductivity was detected in between the CCCL (LSCF) and the Pt interconnect metallized on the AS. It is believed this was caused by EL cracking during temp cycling during sintering due to the CTE mismatch among the layers. The problem might have been exacerbated during cell coating/firing, and cell assembly. However, all first generation electrodes were built in a single batch which allowed no opportunity for fixing the problem. The root cause has been determined and remedy has been incorporated to the next generation. The solution to this problem, when preparing the second generation of cells, is to reduce the CTE of the AS by reducing the relative amount of NiO. The situation could also be improved by using a slightly thicker EL using a more viscous ink. The higher viscosity EL ink will resemble ceramic tape casting slip. Second generation electrodes are been built. Testing of these electrodes will be reported in future publications.

Figure 11. AS3-4 cell assembly, in furnace, connected to external circuit.

CONCLUSIONS

The modeling of anode-supported intermediate-temperature solid oxide fuel cells (ACN-AS-IT-SOFC) and the effect of anode thickness and porosity were accomplished. The simulation study presented was able to identify and compare over-potentials caused by concentration polarization, ohmic polarization, and activation polarization for highly porous anode-supported SOFC. Comparing model simulated results with literature reported experimental data, the exchange current density in the state-of-the-art SOFCs including functional graded layers was found to be four times of that of conventional SOFCs. The effect of anode porosity and thickness on the power density for this type of cell was also studied. At anode porosity around 20%, the power density is sensitive to the thickness of the anode layer. However, when the anode porosity is above 40%, even a thick anode at 1.2 mm may not incur significant drop of the maximum power density. At a fixed anode thickness of 1.0 mm, the anode porosity effect on the power density was studied. Whereas generally a higher maximum power density can be obtained with increasing anode porosity, this effect is less significant when the porosity is above 35%. The results from the model development and derived simulations are of great significance to the optimization design of ACN-AS-IT-SOFC and SOFCs in general.

The feasibility of manufacturing the ACN-AS-IT-SOFC structure was demonstrated. However, validation of the model using corresponding experimental data has not been accomplished yet.

ACKNOWLEDGMENTS

This program was performed under the auspices of the U.S. NAVY under SBIR Phase I Contract N66604-06-C-4640.

REFERENCES

[1] F. Zhao. and A. V. Virkar. Journal of Power Sources, 141, (2005), 79-95.

[2] V. A. C. Haanappel, et al., Journal of Power Sources, 141, (2005), 216-226.

[3] H. Y. Jung, et al., Journal of Power Sources, 155 (2006) 145–151.

[4] H. J. Hwang, et al., Journal of Power Sources. 145, (2005), 243-248.

[5] C-Y Fu. "Deposition of porous LSCF films by EAVD method", Master Thesis, Elec. Theses Heap of NSYSU.

[6] Fuel Cell Materials, Ltd. Worthington. OH. www.fuelcellmaterials.com

[7] Technical Data, Daiicchi Kigenso Kagaku KogyoCo. Ltd., Japan, June 2001.

[8] D.-S. Lee et al., Solid State Ionics. 176, (2005), 33-39.

[9] J. L. Sepulveda. EISG Program. Grant #: 53424A/03-14, PIER. California Energy Commission, May 2005.

[10] P. W. Li, et al., FUELCELL2007-25061, June 18-20, 2007, New York, NY.

[11] P. W. Li, and M. K. Chyu. ASME Journal of Heat Transfer. 127 (2005), 1344-1362.

[12] Wei Guo Wang, and Mogens Mogensen, Solid State Ionics. 176 (2005) 457–462.

[13] Mette Juhl, et al., Journal of Power Sources, 61 (1996) 173-181.

[14] N.T. Hart, et al., Journal of Power Sources. 106 (2002) 42–50.

[15] J. N. Roberts and L. M. Schwartz, Physical review B, 31-9 (1985), 5990-5998.

[16] M. A. Ioannidis, et al., Journal Transport in Porous Media. 29-1 (1997), 61-83.

Electrolytes

HIGH EFFICIENCY LANTHANIDE DOPED CERIA-ZIRCONIA LAYERED ELECTROLYTE FOR SOFC

Juan L. Sepulveda, Sekyung Chang, and Raouf O. Loutfy
Materials and Electrochemical Research (MER) Corporation
Tucson, Arizona 85706, USA

ABSTRACT

This paper describes a novel method to manufacture a higher efficiency, lower cost, long-term stability, planar solid oxide fuel cell (PSOFC) electrolyte to generate electrical power. The SOFC makes use of lanthanide doped ceria-zirconia layered electrolyte. The very high ionic conductivity, functionally graded, hybrid electrolyte was designed to operate at temperature of 800°C or less. Electrolytes tested consisted of yttria stabilized zirconia, scandia stabilized zirconia, scandia and ceria stabilized zirconia, or of a layer of gadolinia doped ceria sandwiched between two layers of scandia stabilized zirconia. Lanthanum strontium manganite and lanthanum strontium cobalt ferrite were used as the cathode while nickel made up the anode. Electrolytes were produced using ceramic tape casting technology and they were tested for SOFC applications using button cells. The advantages of scandia stabilized zirconia over standard yttria stabilized zirconia were demonstrated. Experimentally, a conductivity of 0.0961 S/cm was obtained when using Y_8SZ as electrolyte at 1000°C. The conductivity came down to 0.0128 S/cm operating at 850°C. However, the electrochemical performance obtained for $S_{10}SZ$ was superior with a conductivity of 0.0763 S/cm when operating at 850°C. $S_{10}SZ$ showed better performance than Y_8SZ at lower temperatures of 800-850°C. Operating at lower temperature will result in considerable reduction in production cost. The use of hybrid layered doped electrolytes have the potential to provide even better performance. The feasibility of producing hybrid layered electrolytes was demonstrated and is described in detail.

INTRODUCTION

Solid oxide fuel cells represent a highly efficient power generation/cogeneration source in the United States and worldwide[1-5]. The complete system provides an opportunity to generate electrical energy at overall efficiencies up to 60 % (HHV) with a high temperature thermal by-product that can be used for space heating, industrial processing, or additional electrical generation. They are projected to have low particulate and gaseous emissions and should be readily siteable. SOFC technology does not present the pollution problems associated with conventional fuel or coal based power generators.

This EISG project was awarded under the PIER Subject Area: Renewable Energy Technologies. The project was aimed to prove the feasibility a novel method to manufacture a higher efficiency, overall lower cost, long-term stability, planar solid oxide fuel cell (PSOFC) electrolyte to generate electrical power. PSOFC's generate electrical power by means of clean environmentally friendly alternatives that use renewable energy sources[1-5]. Consuming hydrogen and oxygen (air), and without generating toxic effluents, fuel cell electrical power is a very clean power generating technology. The improved novel fuel cell technology offers several advantages to the State of California such as reduced emissions, higher efficiency, elimination of transmission losses, co-generation capability and enhanced grid reliability. This constitutes very appropriate technology to be utilized in the State of California and in almost all the other remaining States of the Union, using non-polluting technology, by means of a centralized facility or distributed generation facilities. The novel solid electrolyte technology for PSOFC pursued in this project allows the fuel cells to operate at higher efficiency levels resulting in better fuel utilization. The new hybrid electrolyte with a very high ionic conductivity was designed to operate at temperature equal or lower than 800°C.

SOFC TECHNOLOGY

High Efficiency Lanthanide Doped Ceria-Zirconia Layered Electrolyte for SOFC

The novel electrolyte consists either of scandia stabilized zirconia ($Zr_{0.9}Sc_{0.1}O_{1.95-x}$) ($S_{10}SZ$), of scandia and ceria stabilized zirconia ($Zr_{0.89}Sc_{0.10}Ce_{0.01}O_{1.95-x}$) ($S_{10}C_1SZ$), or of a layer of gadolinia doped ceria electrolyte ($Ce_{0.9}Gd_{0.1}O_{1.95-x}$) ($G_{10}DC$) layered between two layers of scandia stabilized zirconia electrolyte ($Zr_{0.9}Sc_{0.1}O_{1.95-x}$) ($S_{10}SZ$) to avoid the reduction of $G_{10}DC$ under the reducing H environment, the main obstacle for the use of ceria doped electrolyte.

A mixture of LSM•$S_{10}SZ$ (lanthanum strontium manganite • scandia-stabilized zirconia) or a mixture of LSCF•$S_{10}SZ$ (lanthanum strontium cobalt ferrite • scandia-stabilized zirconia) was used as cathode while a mixture of NiO•$S_{10}SZ$ made up the anode. The combination of the hybrid solid electrolyte bounded with the cathode at one side, and the anode on the other, constitutes the hybrid single fuel cell or hybrid unit-cell. This innovative fuel cell benefits from the increased ionic conductivity of the electrolyte resulting in higher conversion efficiency, operating at lower temperature so that more affordable materials can be used for interconnects and other components at the same time extending the lifetime of the cell. This novel single cell was designed to increase the cell-stack efficiency from 25% (state-of-the-art yttria stabilized zirconia electrolyte ($Y_8SZ = Zr_{0.92}Y_{0.08}O_{1.96-x}$)) to better than 35% at 800°C without co-generation, allowing an overall complete-system efficiency approaching 65%. As for all SOFC applications, even better performance is obtained at higher cell operating temperatures like 1000°C, but imposing more demanding conditions on metallic interconnects. The higher efficiency derives from the very high ionic conductivity provided by the $S_{10}SZ$, by the $S_{10}C_1SZ$, or by the $G_{10}DC - S_{10}SZ$ hybrid electrolyte as compared to that of standard Y_8SZ. This program demonstrated the feasibility of producing such hybrid single cell and aimed to demonstrate its performance through experimental testing.

The state-of-the-art industrial technology for SOFC applications is centered on the use of yttria stabilized zirconia, with the yttria mol percent varying from 3% to 12%. Most used electrolyte is Y_8SZ (8 m/o Y_2O_3), with an ionic conductivity around 0.1-0.15 S/cm at the typical operating temperature of 1000°C (see Table I). Operating at 1000°C and under a hydrogen/moisture and air corrosive environment imposes very harsh conditions on the metallic interconnect. The only materials currently known to withstand this environment are the high cost high chromium alloys. It is most desirable to reduce the operating temperature to about 800°C, which would in turn allow the use of more affordable ferritic steels for interconnect. However, the ionic conductivity falls to ~0.03 S/cm when operating at 800°C reducing the cell-stack efficiency considerably. New solid oxide electrolyte materials, as those proposed in this project, are needed that can operate at 800°C with an acceptable ionic conductivity.

The relation between the cell terminal voltage, power, and electrolyte conductivity is given by,

$$V = E - \eta_c - \eta_a - IR = E - \eta_c - \eta_a - I/\sigma \qquad (1)$$

and
$$P = VI \qquad (2)$$

where, P is the power, V is the cell terminal voltage, E is the thermodynamic emf (open circuit voltage), η_c is the activation loss at cathode (cathode polarization loss), η_a is the activation loss at anode (anode polarization loss), IR is the ohmic loss at the electrolyte, I is the current, R is the electrolyte resistance, and σ is the electrolyte conductivity. From formula 1 it can be seen that the higher the resistance R is, the higher the ohmic loss will be, and the lower the cell terminal voltage will be. The higher the electrolyte conductivity is, the lower the ohmic loss will be, and the higher the cell terminal voltage will be. Reducing the ohmic loss will increase the electrical efficiency of the cell[6, 7].

Current efforts in planar SOFC technology development concentrate on increasing conversion efficiency, lowering operating temperature, and lowering production cost. Positive improvements are being pursued involving a redesign of electrolyte composition, production technologies, and the use of innovative ancillary materials (interconnect). This EISG project focused on the improvement attained by improving the ionic conductivity derived from the use of $S_{10}SZ$, $G_{10}DC$, and $S_{10}C_1SZ$[8-16]. $G_{10}DC$ exhibits a very high ionic conductivity but it is partially reduced from Ce^{+4} to Ce^{+3} in hydrogen at temperatures above 600°C. The formation of Ce^{+3} ions generates electron holes that make ceria electronically conductive, thus short circuiting the cell. $S_{10}SZ$ also exhibits high ionic conductivity but it is stable in H and O environments at temperatures as high as 1000°C. One of the main technical challenges of this project was to demonstrate the feasibility of using both electrolytes operating in an SOFC environment at 800°C with 25% longer useful life and a fuel cell-stack efficiency comparable to that of the standard Y_8SZ operating at 1000°C. Another technical challenge for this project was to demonstrate the feasibility of producing the novel hybrid electrolyte using plant scale processes and facilities available to the research team.

Table I[8-11] summarizes electrolyte ionic conductivity, σ_i, data gathered from different investigators and suppliers. Ionic conductivity values of recent most advanced electrolytes are compared to that of Y_8SZ, the current SOFC electrolyte market standard. Y_8SZ fuel cells operate at a temperature of around 1000°C. The inherent disadvantage of operating at this high temperature is the highly corrosive condition imposed on interconnects. This EISG project aimed to develop high efficiency SOFC stacks to operate at 800°C which will result in less stringent material demands and longer life for the interconnect ware. From Table 1, it can be concluded that $G_{10}DC$ and $S_{10}SZ$ exhibit ionic conductivity approximately 2-2.5x higher than Y_8SZ at 1000°C. This advantageous performance is even more pronounced at 800°C. More importantly, it can also be concluded from Table I the ionic conductivity obtained for $G_{10}DC$ at 800°C is similar to that of Y_8SZ at 1000°C. A conversion to using $G_{10}DC$ electrolyte and operating at 800°C would result in a cell-stack performance very similar to that being obtained today with Y_8SZ electrolyte operating at 1000°C.

To realize the high σ_i advantages of $G_{10}DC$, the $G_{10}DC$ electrolyte layer has to be kept isolated from hydrogen to avoid reduction occurring at temperatures in excess of 600°C. This is accomplished in the proposed novel design by using insulating layers of hydrogen impermeable $S_{10}SZ$ which are laminated using standard tape lamination technology, and bonded in place during high temperature sintering of the electrolyte.

The most attractive electrolyte configuration resulting in very high ionic conductivity is obtained when the $G_{10}DC$ is sandwiched and protected from reduction from the hydrogen environment with two layers of $S_{10}SZ$. Layering with $S_{10}SZ$ also improves the mechanical strength of the electrolyte. This way, a stable hybrid layered electrolyte structure is obtained. Samarium doped ceria ($S_{15}DC = Ce_{0.85}Sm_{0.15}O_{1.925-x}$) could also be used in place of $G_{10}DC$ resulting in only slighter lower performance. Current tape casting technology allows for casting green tape as thin as 60 μm. This tape would fire to a fired thickness of about 50μm. Typical Y_8SZ tape is cast to 100-500 μm green thickness. Lamination allows to produce Y_8SZ fired substrates as thick as 1500 μm.

Table I. Ionic Conductivity and CTE of Electrolyte Materials for Solid Oxide Fuel Cells.

σ_i (S/cm)

	$G_{10}DC^8$	$S_{15}DC^8$	$G_{10}SZ^9$	$S_{10}SZ^{10}$	Y_8SZ^8
1000°C	0.380	0.263	0.087	0.270 0.300[9]	0.155 0.140[9] 0.150[10]
800°C	0.126	0.100	na	0.090	0.022 0.040[10]
750°C	0.100	0.079	na	0.052	0.013
700°C	0.063	0.050	na	0.020	0.006
650°C	0.039	0.031	na	0.012	0.004
600°C	0.031	0.025	na	0.006	na
550°C	0.020	0.016	na	na	na
CTE 25-1000°C (ppm/°K)	13.4	12.7	na	12.0	10.7

S/cm = Siemens/cm = $\Omega^{-1}cm^{-1}$

Table II[11] compares the total conductivity, σ_t, to the ionic conductivity, σ_i, for $G_{10}DC$ at different partial pressure of oxygen. ($\sigma_t = \sigma_e + \sigma_i$, where σ_e is the electronic conductivity). Table II indicates that σ_e is higher than σ_i for $G_{10}DC$. Using the novel single cell electrolyte concept in this proposal, the increased σ_e is cancelled by using the insulating layers of $S_{10}SZ$ on both sides of the $G_{10}DC$ electrolyte.

Another highly conductive novel electrolyte for SOFC at temperature of 800°C consists of ZrO_2 co-doped with 10 m/o Sc_2O_3 and 1 m/o CeO_2 ($S_{10}C_1SZ$)[16]. This electrolyte did not show any phase transition during heat treatment up to 1550°C, was stable as a cubic phase in all temperature ranges, and showed much higher electrical conductivity than Y_8SZ in the 300-1100°C temperature range. Samples of this electrolyte were also manufactured in this project.

Table II. Total and Ionic Conductivity of $G_{10}DC$ vs. Oxygen Partial Pressure and Temperature.
σ_t (S/cm), σ_i (S/cm), p_{O_2} (atm)

Log p_{O_2}	700°C[11]		800°C[11]		900°C[11]	
	S/cm		S/cm		S/cm	
	σ_t	σ_i	σ_t	σ_i	σ_t	σ_i
-17	0.05	0.028	0.20	0.09	1.60	0.28
-18	0.05	0.03	0.25	0.10	1.70	0.35
-20	0.10	0.037	0.80	0.14	1.80	0.51
-21	0.15	0.04	1.00	0.21	2.21	0.60
-22	0.20	0.043	1.10	0.30	2.41	0.67
-23	0.20	0.05	1.35	0.38	na	na
-24	0.40	0.07	1.87	0.47	na	na
-25	0.50	0.10	2.10	0.56	na	na

The cathode and anode can be deposited using conventional thick-film screen printing technology or a computer controlled rapid prototyping coater to build the planar SOFC single cell. The prototype hybrid button cells built in this project were electrolyte supported as shown in Figure 1. Currently, electrolyte-supported cells use 300 to 400 μm thick electrolyte. The target electrolyte thickness of this project was around 300 μm to lower the resistance loss and to lower cost. The addition of $G_{10}DC$ in the center of the $S_{10}SZ$ electrolyte results in improved ionic conductivity at 800°C. Cell-stack performance will improve from the current 25% efficiency to 35% efficiency or larger[4, 18]. A small amount of $S_{10}SZ$ or Y_8SZ was used for the anode and cathode formulation to mitigate the problem associated with thermal expansion mismatch especially between the electrodes and the electrolyte, to promote electrochemical activity (increased active sites), and to promote porosity[4, 19, 20].

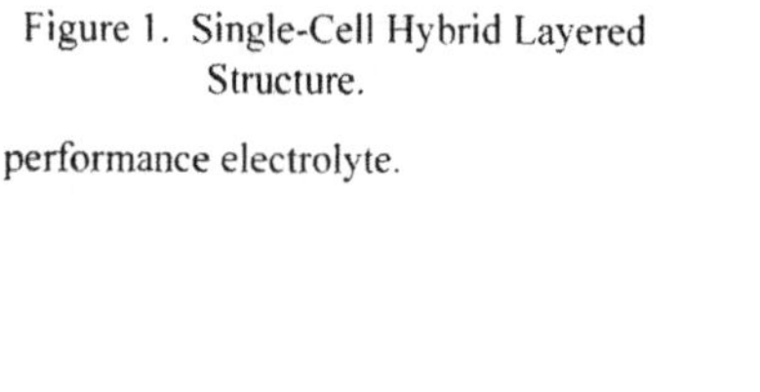

The main objectives of this project were to demonstrate manufacturing feasibility and perform the preliminary testing to demonstrate the effectiveness of the layered structure shown in Figure 1 as a high performance electrolyte.

Figure 1. Single-Cell Hybrid Layered Structure.

EXPERIMENTAL

The first approach to build the layered structure consisted of using conventional tape casting technology[17] to produce $S_{10}SZ$, $G_{10}DC$, and $S_{10}C_1SZ$ green tape. After tape casting and drying, the tape was laminated under pressure and temperature using standard lamination procedures and laser cut to comply with electrolyte geometry and sintered. An example is shown in Figure 2. Another alternative

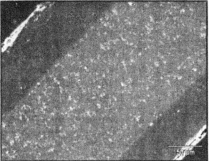

Figure 2. Cross SEM micrograph of 4 layered:
1 $S_{10}SZ$ – 2 $G_{10}DC$ - 1 $S_{10}SZ$ electrolyte
polished cross section at 400 X.

of producing hybrid electrolyte was attained by using 10 m/o Sc_2O_3, 1 m/o CeO_2, co-doped $ZrO2$ powder ($S_{10}C_1SZ$). This electrolyte presents several advantages such as higher conductivity than Y_8SZ, eliminates phase transitions, and stabilizes the zirconia structure at operating temperatures of up to $1550°C$[16]. Yttria-stabilized zirconia (Y_8SZ) powder was also used as electrolyte to define base line performance. A mixture of toluene and ethanol in 60%:40% by weight ratio was the solvent used during tape casting. Polyvinyl butyral (B-98), santicizer S-160, and fish oil (Z-3) were used as binder, plasticizer, and dispersant respectively. In general, the rheological behavior of all prepared slurries was very similar to that observed previously for standard Y_8SZ. Slurry viscosity was kept 3000-5000 cp in all casts.

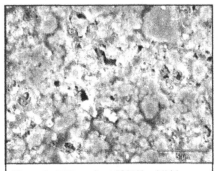

Figure 3. NiO anode at 6000X exhibiting porosity to facilitate hydrogen diffusion.

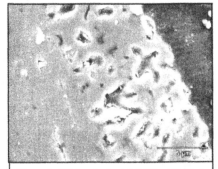

Figure 4. LSM cathode at 6000X exhibiting porosity to facilitate oxygen diffusion.

The anode material comprised a mixture of Y_8SZ or $S_{10}SZ$ and high purity NiO in 50:50% volume ratio, while the cathode consisted of lanthanum strontium manganite (LSM) or lanthanum strontium cobalt ferrite (LSFC) and Y_8SZ or $S_{10}SZ$ in 70:30% volume ratio. Typical microstructures are shown in Figure 3 and Figure 4.

Anode formulations used for standard Y8SZ electrolytes contained Y_8SZ while anode formulations used for $S_{10}SZ$ electrolytes contained $S_{10}SZ$. Cathode formulations used for standard Y_8SZ electrolytes were produced using LSM and contained Y_8SZ. Cathode formulations used for $S_{10}SZ$ electrolytes were produced using LSCF and contained $S_{10}SZ$. LSM cathode formulations are appropriate when operating SOFC at 1000°C as for standard Y_8SZ electrolyte. Cathode formulations containing LSCF are more appropriate for SOFC application operating at 600°C to 800°C. Other cathode formulations for fuel cells operating at these lower temperatures, which were not tested in this project include: lanthanum strontium ferrite (LSF), lanthanum strontium manganese ferrite (LSMF), praseodymium strontium manganite (PSM), and praseodymium strontium manganese ferrite (PSMF).

Tape casting of electrolyte slurry was accomplished using either a 4 ft batch TAM 165 tape caster or 60' long A. J. Carsten continuous tape caster. Green tape sheets were laminated in a hydropress vessel to produce single or hybrid chemistry electrolytes.

Alternatively, the green $S_{10}SZ$ layers were deposited on previously sintered $G_{10}DC$ electrolyte using a computer controlled rapid prototyping coater and using the same $S_{10}SZ$ ceramic slurry. This technique allowed for better control of the $G_{10}DC$ fired microstructure and better control of the thickness of the $S_{10}SZ$ layers. After drying, the $S_{10}SZ$ layers were fired at 1450°C for two hours.

Drying of the green tape during casting was accomplished by blowing filtered air through the tape-casting machine. The solvent loaded discard air was sent to the incinerator (at temperature in excess of 1200°C) unit prior to disposing of the residual effluent. The apparent tape bulk density was determined using the weight of a specimen and its corresponding measured volume.

From all the results gathered during this project it can be concluded that all the electrolyte powder formulations were successfully cast with the exception of nano-sized electrolytes which being this fine were not appropriate for casting. Micronized powders were more appropriate for casting. The most appropriate viscosity at which all casts were performed was around 3000 cp.

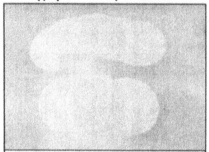

Figure 5. 3-Layer electrolytes 1.9" and 2.2".

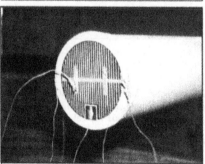

Figure 6. Bonded button cell. Cathode provided with wrap-around metallization. Brazed Pt interconnects pointing to the front of the cell.

To manufacture hybrid layered electrolytes the $G_{10}DC$ and $S_{10}SZ$ layers were laminated and co-fired in one single unit firing operation in which binder removal and sintering are accomplished in a single process. Experiments were carried out to optimize the heating rate to ensure that no cracks, warp, or other defects develop during the firing. Sintering was carried out at 1250-1500°C in air for 2-4 hrs. The shrinkage, fired density, and flatness were measured after firing. SEM micro-structure analyses shown in Figure 2 revealed the multilayer microstructure accomplished.

It was concluded all stabilized zirconia electrolytes densified very well. Values close to 100% theoretical density were obtained. However, the $G_{10}DC$ co-fired, layered, electrolyte showed some 2-4% residual porosity. This porosity resulted in weaker electrolyte that mechanically failed during electrochemical testing. Production of layered hybrid electrolyte was switched to using a higher density sintered $G_{10}DC$ substrate that was coated with $S_{10}SZ$ using the computer controlled coater. The electrolytes were fired to 1450°C for two hours after coating.

All electrolytes produced in this project were flat with camber of less than 0.003"/" as determined by the parallel plates procedure. Flat firing was also used in some cases to improve flatness. No problems were encountered during flat firing.

The anode and cathode formulations were tailored to produce thermal expansion close to that of the tape cast electrolyte. The anode fired density was aimed to reach approximately 60% to 70% of theoretical while the cathode was also around 60% to 70% theoretical density. A slurry of the anode material was produced by mixing $S_{10}SZ$ (50 vol.%) with nickel oxide. The anode powder mix was wet-milled in terpineol using a Nalgene lined ball-mill. Binder consisted of ethyl cellulose and was added to the mill. The cathode was prepared from a mixture of LSM or LSCF and $S_{10}SZ$ (~30% vol). The anode and the cathode slurries were coated onto the fired electrolytes one side at a time using a computer controlled rapid prototyping coater. Firing of the anode and cathode layers were accomplished in air atmosphere at 1250°C for 2 hr and 1100°C for 2 hr respectively. The Pt interconnect was also printed under computer control and fired in air at 1050°C for 10 minutes. Heraeus frittless Pt paste was used for this purpose. Pt wires were bonded to the electrolytes using the same Pt paste. The unit-cell flatness was 0.003 in/in or better. Examples of electrolyte produced are shown in Figures 5. Figure 6 shows an example of the Pt metallization with a fish bone pattern which was printed on both anode and cathode side as interconnect.

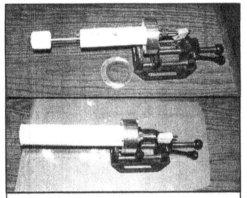

Figure 7. Button cell inner components and assembled ready for insertion into the furnace.

Electrolyte and electrode compositions were evaluated using 1.90" diameter disk shaped cells (button cells) in a bench top apparatus. As shown in Figure 6 and Figure 7, the button cells were bonded to the end of a zirconia tube, the inner face exposed to hydrogen (anode) and the outer side exposed to air (cathode). The button cell assembly shown in Figure 7 was placed inside a clamshell heater. Hydrogen was fed through a 1/8 inch diameter tube and impinged on the internal face of the button cell. A porous zirconia firebrick block was used to insulate the hot end from the cold end of the cell. Steady state DC voltage sweeps (low rate sweeps) were used to evaluate performance of the button cells at temperatures between 600°C and 1000°C. using humidified hydrogen and air. Current was measured as function of voltage to generate polarization curves or V-I profiles. Characteristics of the profile can be interpreted to give electrolyte resistance. exchange current losses, combustion reaction consumptions, internal electrode resistance and 3-point contact resistance. A resistive load or a highly regulated DC power supply integrated with a controller and data logger were used to polarize the cell[22, 23]. An air-side reference electrode was incorporated to measure the cathode overpotential. The performance of button cells was characterized over the voltage range from Nernst (open circuit voltage) (~1.1 V) to ~0.4V at a fixed temperature. From V-I curves obtained at different temperatures, Arrhenius plots were prepared to determine the activation energy. Testing was also carried out independently at a customer facility that requested to remain anonymous. These tests consisted in determining the cell voltage obtained at constant current for extended periods of time. These tests were also performed for the purpose of validating results.

EXPERIMENTAL RESULTS

The performance base line was determined using Y_8SZ. The polarization curves obtained at different temperatures are shown in Figure 8. From the slope of the curves in the ohmic region, the conductivity was calculated at different temperatures as shown in Table III. The corresponding Arrhenius plot is shown in Figure 9. The dependence of the conductivity on inverse absolute temperature was linear as indicated by the Arrhenius plot in a semi-log chart shown in Figure 9.

Cell performance was sensitive to the contact between the electrodes and the current collector[21]. The fish bone pattern shown in Figure 6 provided the best results.

Table III. Y₈SZ Electrolyte Conductivity at Different Temperatures.

Temperature, (°C)	Conductivity, (S/cm)
800	0.0059
850	0.0128
900	0.0240
950	0.0471
1000	0.0961

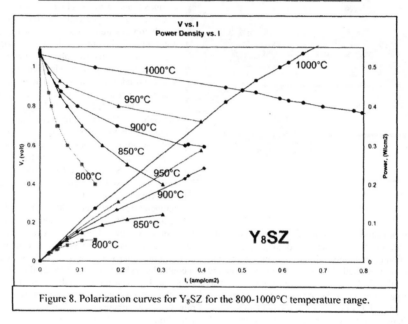

Figure 8. Polarization curves for Y₈SZ for the 800-1000°C temperature range.

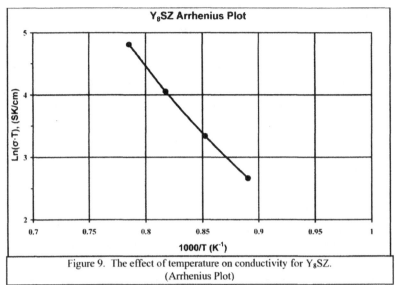

Figure 9. The effect of temperature on conductivity for Y_8SZ.
(Arrhenius Plot)

Figure 10 compares the effect of using Ag-Pd as full face anode/cathode coverage interconnect over a Pt simple pattern, to using the Pt pattern exclusively. No apparent advantage was observed. The temperature had to be kept at 800°C maximum to avoid interconnect melting.

Figure 11 shows the polarization curve obtained for $S_{10}SZ$ at 850°C as compared to standard

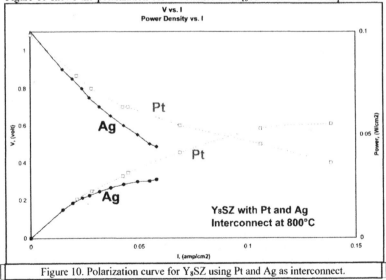

Figure 10. Polarization curve for Y_8SZ using Pt and Ag as interconnect.

Y_8SZ at 850°C and 1000°C. The electrochemical performance of $S_{10}SZ$ at 850°C is superior to that of Y_8SZ at 850°C. From the slope of the curve in the ohmic region a conductivity of 0.0763 (S/cm) was obtained. It can be concluded $S_{10}SZ$ electrolyte is a good candidate for SOFC applications operating in the 800-850°C temperature range.

Figure 12 and Figure 13 show fuel cell performance under steady conditions for extended periods of time. Figure 12 shows the voltage and current generated for a period of approximately 2 hours when Ag-Pd was used as interconnect. Figure 13 exhibits voltage generated at a constant current density of 0.45 amp/cm², at a temperature of 1000°C, during 30 days. Figure 13 results were generated at an external customer. Standard Y_8SZ performance is compared to $S_{10}SZ$ in Figure 13.

The advantageous performance of $S_{10}SZ$ over Y_8SZ was estimated to be approximately 7%. The cell performance was maintained for 30 days with no apparent degradation. These results confirm the advantageous performance of $S_{10}SZ$ over Y_8SZ observed from button cell testing although at a higher temperature.

All fuel cell testing attempted on the laminated layered hybrid electrolyte has not been successful yet. These parts did not have sufficient strength to endure the sample preparation needed prior to testing. Parts mechanically failed after anode/cathode attachment or after interconnect attachment. A different process was tried to produce these parts using a denser $G_{10}DC$ core layer that is sintered in a first firing operation to higher density, coating the fired $G_{10}DC$ with $S_{10}SZ$, and firing the layered structure in a second firing operation at 1450°C for two hours. Experimental runs have not been successful yet.

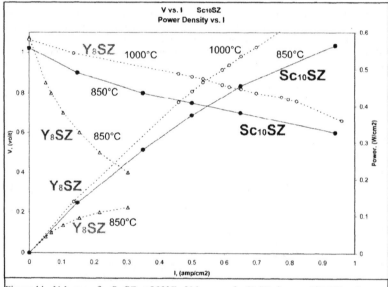

Figure 11. V-I curve for $S_{10}SZ$ at 850°C. V-I curves for Y_8SZ shown at 850°C and 1000°C for comparison purposes.

CONCLUSIONS

This project demonstrated the feasibility of producing advanced hybrid electrolyte for more efficient solid oxide fuel cell applications using tape casting technology. The electrochemical performance obtained for Y_8SZ (conductivity of 0.0961 S/cm at 1000°C) is in agreement with values reported by other researchers. The electrochemical performance obtained for $S_{10}SZ$ (conductivity of 0.0763 S/cm) at 850°C is in agreement with values reported by other researchers and it was superior to that observed from Y_8SZ at the same temperature (conductivity of 0.0128 S/cm). The advantageous performance of $S_{10}SZ$ over Y_8SZ was also verified after running extended cell operation at a customer site. An increase in voltage at constant current density of 7% was calculated.

$S_{10}SZ$ showed better performance than Y_8SZ at lower temperatures of 800-850°C. Electrolytes that can operate at this lower temperature range will result in considerable reduction in production cost estimated at 70%. It was estimated that $S_{10}SZ$ at 900°C would equal the performance of Y_8SZ at 1000°C.

The use of hybrid layered or co-doped electrolytes has the potential to provide even better

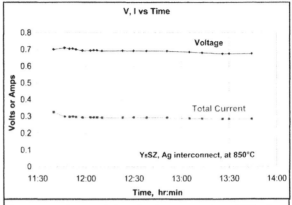

Figure 12. Voltage variation with time for Y_8SZ, at 850°C, when using Ag-Pd as interconnect.

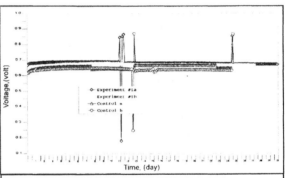

Figure 13. Voltage variation with time at a fixed current density of 0.45 amp/cm². Standard Y_8SZ (control a and b) compared to $S_{10}SZ$ experimental #1a and #1b at 1000°C.

performance. This project did not succeed in demonstrating this behavior within the allowed project time frame. Although no major problems were found developing the different tape casting/firing processes for the new electrolytes, the strength of the fired layered electrolytes was not sufficient to allow further electrochemical testing. The strength of these electrolytes has to improve either by small changes in composition, by increasing the relative thickness, by better matching the thermal expansion of the different layers, or by improving the quality of the ceramics. Some researchers have used interlayers to get a better thermal expansion match[21].

The anode/cathode formulations varied for the different electrolytes. An extensive optimization is required in this area which was not possible during the duration of this project. It could be concluded that electrochemical performance could increase significantly by using and adjusting the proper anode and cathode formulation to each electrolyte, this way reducing the polarization losses. This becomes a very important design feature when operating at the lower temperatures (~800°C) attempted here. Important factors in these formulations are not only the metallic material used but also the ceramic material used to produce the porous cermets. Ceramic materials that exhibit maximum electrochemical reactivity under fuel cell operating conditions are most desirable.

The contact between the interconnects and the anode or the cathode was another very important factor in the performance of the cell. By maximizing the contact area optimum results will be obtained. Time permitting, more optimization could have been done in this regard in this project. The challenge is to attain maximum contact area and at the same time allow gas transport while minimizing the thermal mismatch between the electrolyte and the metallic interconnect.

The use of button cells to demonstrate fuel cell performance was preferred to complex impedance measurements since button cells represent more closely operating condition and materials of construction for SOFC. However, button cell techniques require much more efforts to produce results.

ACKNOWLEDGMENTS

This program was performed under the auspices of the California Energy Commission under EISG Contract 53424A/03-14. All experimental work was performed at Intertec, Tucson, Arizona and RP Circuits, Tucson, Arizona, facilities.

REFERENCES

[1] 2002 Fuel Cell Seminar, Palm Springs, California, Nov. 18-21, 2002.

[2] Q. M. Nguyen et al., Sci. and Tech. of Cer. Fuel Cells, Amsterdam, Elsevier Science, 1995.

[3] Proc. of the Third Int. Sym. on Solid Oxide Fuel Cells, Eds. S. C. Singhal and H. Iwahara, The Electrochemical Society, Inc., Pennington, NJ, 1993.

[4] Fuel Cell Handbook Sixth Edition. DOE/NETL-2002/1179

[5] President George W. Bush, "State of the Union Address", January 28, 2003.

[6] Mann et al., 2002 Fuel Cell Seminar, November 18-21, 2002, Palm Springs, CA

[7] Fowler et al., J. of Power Sources 106, 2002, pp. 274-283

[8] Fuel Cell Materials, Ltd., Worthington, OH. www.fuelcellmaterials.com

[9] Y. Arachi et al., Solid State Ionics 121 (1999) pp. 133-139

[10] Technical Data, Daiichi Kigenso Kagaku Kogyo Co. Ltd, Japan, June 2001

[11] S. Wang et al., J. of Electrochemical Society, 147 (10), 2000, pp. 3606-3609

[12] J. Van Herle, et al., Proc. 4th Int. Symp. on SOFC, Yokohama, Japan 1995.

[13] N. Sammes et al., Denki Kagaku, No. 6, 1996.

[14] H. Yahiro, et al., Solid State Ionics 36 (1989) 71-75

[15] K. Eguchi, et al., Solid State Ionics 52 (1992) 165-172.

[16] D.-S. Lee et al., Solid State Ionics, 176 (2005) 33-39.

[17] J. S. Reed, "Principles of Ceramic Processing", John Wiley, New York, 1988

[18] P. N. Kumta, et al., Electrochemical Materials and Devices", ACerS, 1999.

[19] D.Y. Wang, D.S. Park, J. Griffith and A.S. Nowick. Solid State Ionics, 2, 95 (1981)

[20] R.Gerhard-Anderson, and A.S. Nowick, Solid State Ionics, 5, 547 (1981)

[21] S. L. Swartz, et al., 2002 Fuel Cell Seminar, November 18-21, 2002, Palm Springs, CA.

[22] M. Smith, et al., Fuel Cell Magazine, April/May 2005, 26-31.

[23] J. Niemann, Fuel Cell Magazine, April/May 2005, 32-37.

[24] S. P. Jiang et al., Solid State Ionics 8914 (2003) 1-12

OXYGEN ION CONDUCTANCE IN EPITAXIALLY GROWN THIN FILM ELECTROLYTES

S. Thevuthasan[*,1], Z. Yu[2], S. Kuchibhatla[1,3], L.V. Saraf[1], O. A. Marina[1], V. Shutthanandan[1], P. Nachimuthu[1] and C. M. Wang[1]

[1]Pacific Northwest National Laboratory, Richland, WA-99352, USA
[2]Department of Chemistry, Nanjing Normal University, Nanjing 210097, China
[3]Advanced Materials Processing and Analysis Center, &Nanoscience and Technology Center, University of Central Florida, Orlando, FL-32816, USA

ABSTRACT:

This paper briefly summarizes the results from a project aimed to develop an understanding of oxygen ionic transport processes in highly oriented thin film oxide materials to enable the design of new types of electrolyte materials for solid state electrochemical devices. We have used oxygen-plasma-assisted molecular beam epitaxy (OPA-MBE) to grow highly oriented doped ceria, zriconia thin films on single crystal c-Al_2O_3 along with multilayered hetero-structures. The influence of dopant concentration, interfaces, defects and crystalline quality on oxygen ionic conductivity has been critically analyzed using various surface and bulk sensitive capabilities. Although, preferred (111) orientation was preserved in high quality samaria doped ceria films up to a 10 atom% Sm doping, the films started to show polycrystalline features for higher Sm doping. Maximum conductivity was obtained for 5 atom% Sm doping in ceria. In the case of gadolinia doped ceria/zirconia multi-layer films, total conductivity was found to increase with the increasing number of layers.

INTRODUCTION

The operation of all solid-state electrochemical devices including batteries, sensors, fuel cells and oxygen pumps is essentially built on ion conduction in solid electrolytes between two electrodes. In case of a solid oxide fuel cell (SOFC) an oxide electrolyte membrane, that is typically yttria-stabilized zirconia (YSZ), is provided with two electrodes exposed to air and hydrogen. The difference in oxygen activity at the electrodes drives oxide ions through the electrolyte from the air electrode (cathode) to the fuel electrode (anode), where they react with hydrogen producing water and releasing the electrons. The operation temperature of electrochemical devices is governed mainly by the electrolyte and electrode resistances, which is typically in the range of 1000°C or above for electrolytes based on YSZ. However, high operating temperatures significantly restrict the choice of materials for other structural components such as interconnects, manifolds and cell housing to relatively expensive materials. Although decreasing SOFC operating temperature provides means to utilize cheaper materials, the cell resistance increases significantly with decreasing temperature and limits SOFC performance to higher temperatures. Reduction of electrolyte thickness will reduce the resistance in the electrolyte and, in turn, this will help in achieving higher conductivities. Development of electrolyte materials with higher oxygen ion conductance at lower temperatures would further contribute to an increase in the efficiency and lifetime of many electrochemical devices, including SOFCs.

*Contact Author: theva@pnl.gov

Oxygen Ion Conductance in Epitaxially Grown Thin Film Electrolytes

Driven by the need for alternate energy resources and the thrust on hydrogen economy, fundamental understanding of oxygen transport behavior in the solid oxide electrolytes has been a major interest of research community for decades [1, 2]. To increase the ionic conductivity of the electrolyte and, thereby, to reduce the operating temperature of the electrochemical devices, several strategies are currently being explored. The prevalent approach to increase the conductivity is based on the incorporation of different dopants, which preferentially increase the concentration of certain types of lattice defects in the electrolytes [3, 4]. This strategy shows promising results, but not always is sufficient. Consequently, alternative materials are being searched for, and cerium oxide doped with a divalent or trivalent oxide (mostly gadolinia or samaria) has been found to exhibit higher ionic conductivity compared to YSZ at lower temperatures around 1075 K with a lower activation energy [5-12]. Considerable electronic conductivity of ceria when exposed to the reducing atmospheres at temperatures above 825 K leads to voltage losses. In addition to the doped ceria materials, recent studies demonstrate that scandia doped zirconia is another promising material that shows higher ionic conductivity in comparison to YSZ [13-18]. A huge database exists in the literature based on the powder processed materials which suffer various inherent disadvantages sprouted from the chemical synthesis route. In addition to this, the polycrystalline nature of the materials often results in complicated impedance data that hinders a comprehensive understanding of the material behavior.

Hence, we have synthesized epitaxial thin films of doped ceria on single crystal Al_2O_3 substrates using OPA-MBE. In addition to the high quality epitaxial films, we have also prepared nanoscale multi-layer hetero-structures of doped ceria/zirconia to explore the conductivity mechanisms. All the films were characterized using several surface and bulk sensitive techniques. The films showed high crystalline quality and the maximum conductivity was obtained for 5 atom % Sm [19] doping in ceria. In the case of gadolinia doped ceria/zirconia multi-layer films, total conductivity is found to increase with the increasing number of layers [20].

EXPERIMENTAL DETAILS

The films were grown in a dual-chamber ultrahigh vacuum (UHV) system [21] equipped with an electron cyclotron resonance (ECR) oxygen plasma source. Ce and Zr sources (both 99.98% purity) were evaporated from separate electron beam sources and Sm, Gd (99.98% purity) were evaporated from effusion cells. Al_2O_3 (0001) single crystal substrates were ultrasonically cleaned in acetone and methanol prior to insertion into the dual-chamber UHV system through a load lock. Inside the MBE chamber, the substrate surfaces were cleaned at 875 K for 10 minutes in activated oxygen from the ECR plasma source at an oxygen partial pressure of $\sim 2.0 \times 10^{-5}$ Torr. All the films were grown in an oxygen partial pressure of $\sim 2.0 \times 10^{-5}$ Torr and at a substrate temperature of 650^0C. The growth rates of the dopants (samaria and gadalonia) and the matrix (ceria and zirconia) were monitored by quartz crystal oscillators (QCOs). Optimum growth rates were established by comparing the data from thickness monitors, XPS depth profile measurement and RBS analysis. In the case of multi-layer films, the gadolinia-doped epitaxial structures of ceria and zirconia layers were grown through alternate evaporation of Ce and Zr metals in oxygen plasma. The optimum conditions for the growth of high-quality single crystal ceria films, highly oriented GDC and GSZ layers and their characterization have been published previously [20].

The film growth was monitored using *in situ* reflection high energy electron diffraction (RHEED). The epitaxial growth of these films was confirmed by x-ray diffraction (XRD) and transmission electron microscopy (TEM). X-ray photoelectron spectroscopy (XPS) measurements were performed to determine the oxidation state of elements and the elemental distribution throughout the film using sputter depth profiling. Rutherford backscattering spectrometry (RBS) and TEM were used to characterize the crystalline quality and the nature of the interfaces. The total conductivity measurements were carried out using the four-probe Van der Pauw technique [22]. Four identical Au electrodes-terminals were spring-loaded to the ceria film. Probe tips were polished to reduce the Ohmic contact resistance. The separation between the probes was kept nearly even and significantly larger than the thickness of the film. Characteristic resistances R_1, R_2, R_3, R_4, associated with the corresponding terminals 1, 2, 3 and 4 were measured in the temperature range 300-900^{0}C using an ac electrochemical impedance analyzer. Each R was found as the distance from the low frequency intercept and the origin (high frequency intercept) of the real part of the corresponding impedance spectra. The difference between all four resistances was within 10-15%. Thus, the assumption for symmetrical geometry was made and the electrical conductivity σ, was calculated using the van der Pauw equation:

$$1/\sigma = \pi d R_s/\ln(2) \tag{1}$$

where R_s is sheet resistance, d is the thickness of the film, $\pi = 3.14$. It has been shown that the use of the ac impedance spectroscopy technique may cause the parasitic inductive effects in both low and high frequencies [22]. Indeed; small inductive loops were periodically observed in the high frequency part. However, the size of these loops was negligible compared to the rest of the impedance spectra and those contributions were not taken into account.

RESULTS AND DISCUSSION

(1) SAMARIA DOPED CERIA THIN FILMS

The structural characteristics of samaria-doped ceria films were analyzed by x-ray diffraction (XRD, Phillips X'pert θ-2θ diffractometer) and high-resolution transmission electron microscopy (HRTEM, JEOL 2010 microscope). Rutherford backscattering spectrometry (RBS) measurements were carried out to investigate the interfacial diffusion along with the film thickness and stoichiometery. XRD measurements (not shown) indicated that the film was predominantly oriented with the CeO_2 (111) plane parallel to the substrate basal plane [19].

Selected area electron diffraction measurements were carried out on the cross-sectional TEM specimen. In this diffraction pattern (not shown), the zone axes for the fluorite structured samaria-doped CeO_2 film and the sapphire substrate are [110] and [11-20], respectively [19]. Selected area electron diffraction pattern shows the orientation relationship: CeO_2 [110]//Al_2O_3[11-20] and CeO_2 (111)//Al_2O_3(0001), which is consistent with the XRD results. No apparent hkl reflections related to Sm-O phase were observed. At Sm concentrations above 10 atom%, films suffered a loss of orientation and transformed into polycrystalline samaria-doped ceria phase. This might be a result of a structural instability beyond a critical dopant concentration. XPS depth profile gives an excellent overview of atomic elemental concentrations of Ce, Sm, O and Al as a function of film depth. The qualitative variation in dopants and associated oxygen vacancies can be realized and applied in conjunction with film conductivity. Representative XPS depth profile for the 5 atom% samaria-doped CeO_2 film is presented in Fig. 1 [19]. A uniform concentration of samaria as a function of film depth was observed and the stoichiometry of the film was determined to be $Ce_{0.90}Sm_{0.10}O_{1.86}$. In addition, fairly sharp interface between samaria doped CeO_2 and Al_2O_3 is also realized from the spectra indicating minimum diffusion.

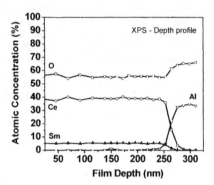

Fig. 1: The XPS depth profile of $Ce_{0.90}Sm_{0.10}O_{2-\delta}$ film on c-Al_2O_3. These measurements were done after conductivity measurements. A uniform Sm concentration of 5 atom% throughout the film can be seen in the data [19].

Conductivity of samaria-doped ceria films with different amounts of samarium dopant was measured in the temperature range of 250-900 °C (Figure 2). As seen, the 5 atom% samaria-doped CeO_2 sample exhibited the highest conductivity throughout the measured temperature range. The activation energy appears to be constant over the temperature range of conductivity measurements for each Sm concentration.

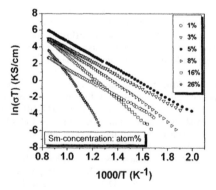

Fig. 2: Arrhenius plots for samaria doped ceria films with various Sm atomic concentrations (x) in the temperature range 250-900 °C [19]

Total conductivity values determined as a function of Sm concentration from Fig. 2 at different temperatures are presented in Fig. 3.Total conductivity increases as a function of Sm concentration, peaks around 5 atom % Sm and decreases for higher Sm concentrations. The reduction in the conductivity beyond an optimum concentration of 5 atom% might be a result of possible loss of orientation in the films which eventually become polycrystalline beyond 10 atom% Sm concentration. The conductivity measurements through impedance spectroscopy consist of both ionic and electronic contributions. However, in case of doped ceria system it can be safely assumed that the electronic component is negligible as the Ce is in its highest oxidation state (4+) and reduction of the same is almost negligible in air under the measurement conditions. Also, the conductivity maxima is observed at significantly lower dopant concentration, relative to reported values of approximately 6-7 atom% , in case of samaria doped ceria samples. This can be ascribed to the oriented nature of MBE grown films in contrast to purely polycrystalline doped samples in the literature. The ordered arrangement of oxygen vacancies in an oriented single crystal contribute to significant increase in the oxygen ionic conductivity through the material. In contrary, increase in grain boundaries in polycrystalline materials show blocking effect to the ionic transport across them due to oxygen vacancy depletion in grain boundaries [23]. As such, in the high quality well oriented films in this study, a good orientation with ordered oxygen vacancies would have ultimately contributed to a good conductivity. The activation energy E_a, derived from the slope of the Arrhenius plot for the 5 atom% Sm doped ceria sample is observed at 0.6 eV. Naturally, lower activation energy refers to lower barrier for the oxygen transport.

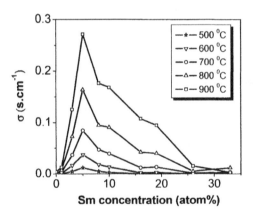

Fig. 3: The conductivity of samaria doped ceria thin films as a function of Sm concentration for the temperature range of 500-900 $^{\circ}$C. At all measurement temperatures, highest conductivity is seen at 5 atom % Sm [19].

It is clear from the aforementioned discussion that the high quality oriented samaria doped ceria films grown by OPA-MBE show significantly higher conductivities at lower activation energy in the low and intermediate temperature range. The reduction in the conductivity data beyond an optimum concentration of dopant is attributed to the crystalline instability leading to the formation of polycrystalline structures further contributing to the vacancy scattering through grain boundaries.

(2) GADOLINIA DOPED CERIA/ZIRCONIA MULTI-LAYERS

The growth of multi-layer structures was monitored by the *in-situ* RHEED. The RHEED streaks indicate the epitaxial nature of the films (not shown here). The XRD results indicate that the multi-layers consist of more than a single phase. The XRD pattern obtained for the hetero layers of gadolinia doped ceria (GDC) and gadolinia stabilized zirconia (GSZ) on the $Al_2O_3(0001)$ substrate is shown in Fig. 4. In addition to the substrate peaks, distinct peaks corresponding to GDC and GSZ films confirm the cubic fluorite structure. In addition, the GDC portion of the film mostly consists of a single phase with (111) orientation. The GSZ portion of the film is highly oriented along (111) direction, with a minor (002) orientation.

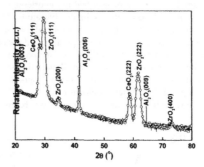

Fig. 4: XRD data for hetero-layers of Gd_2O_3-doped CeO_2 and ZrO_2 thin films on Al_2O_3 (0001).

RBS (Rutherford backscattering spectrometry) experiments along with the XPS depth profile measurements were performed in order to investigate the interfacial properties and elemental interdiffusion phenomena for these layered structures. A representative RBS spectrum along with simulated results using SIMNRA from a 4 layer film is shown in Fig. 5. These results also indicate that, within the depth resolution of about 15-20 nm, there was no elemental interdiffusion at the interfaces between the substrate and the film. The simulated results with two GDC and two GSZ layers agree well with the experimental data. Thickness of each GDC and GSZ layer appear to be 35 nm and 44 nm, respectively and the total thickness of the film is approximately 155 nm. The total thickness was kept constant while the individual layer thickness was varied between one, two, four, eight, ten and sixteen layer films. The average gadolinia concentration in both ceria and zirconia layers appears to be ~12 atom %.

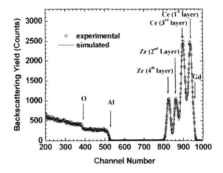

Fig 5: RBS spectrum for epitaxially grown hetero layers of Gd_2O_3-doped CeO_2 and ZrO_2 thin film (4 layers) on Al_2O_3 (0001). Incident energy of the He^+ beam was 2.04 MeV and the scattering angle was 150°.

Although all the layers were grown under the same deposition conditions, the gadolinia concentration appears to be different between the ceria and zirconia films as shown in the results from XPS depth profiles and energy dispersive spectroscopy (EDS) measurements (Fig. 6). The comparison of gadolinium and oxygen XPS depth profiles with those from EDS using scanning transmission electron microscopy (STEM) is presented in Fig. 6. These results indicate that the defect concentrations (Gd incorporation and O vacancies) are not dictated by the growth conditions. Instead, the energetics associated with the defects and strain relaxation enhances the concentration of gadolinia in ceria and, as a result, oxygen vacancies are higher in the ceria layers compared to zirconia layers. These elemental distribution features appear to be rather stable even following high temperature annealing of the specimens. This is demonstrated in Fig. 7, which compares XPS depth profile results from as deposited and annealed samples. The annealing treatment was carried at 1175 K for 10 hours to test the stability of the layered structures.

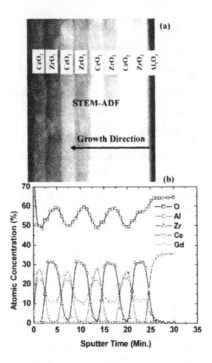

Fig. 6: XPS depth profiles from an eight-layered Gd$_2$O$_3$-doped CeO$_2$ and ZrO$_2$ thin film are compared with the scanning transmission electron microscopy energy dispersive spectroscopy (STEM-EDS) elemental distribution.

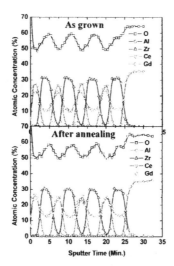

Fig. 7: XPS depth profiles from eight-layered Gd_2O_3-doped CeO_2 and ZrO_2 thin film in top) as grown and bottom) annealed film at 1175 K for 10 hours in air.

The cross-sectional TEM micrographs the 2, 4, 8, and 16 layer films are presented in Fig. 8. The inset in each image is the selected area electron diffraction pattern (SAEDP), which reveals epitaxial orientation relationship of the layers. The SAEDP of the films mainly show that both gadolinia doped ceria and gadolinia-stabilized zirconia layers are single crystals with different lattice parameters as clearly revealed by the separation of the high order diffraction spots and they maintain a well defined epitaxial orientation relationship. The lattice mismatch, δ, between the CeO_2 and ZrO_2 layers can be defined as: $\delta = (a_{Ce}-a_{Zr})/a_{Zr}$ [25] where a_{Ce} and a_{Zr} are the in-plane lattice parameters of the CeO_2 and the ZrO_2, respectively. δ can be determined from the separation of the spots on the SAEDP, giving a value of ~ 5.7%. Using the lattice parameters of 0.541 nm for c-CeO_2 [26] and 0.514 nm for c-ZrO_2 [27], a δ value of ~ 5.3% is obtained. It should be noted that determination of lattice constant by selected area electron diffraction method gives an uncertainty of 2%. In addition, the CeO_2 and ZrO_2 layers maintain a very well defined epitaxial orientation relationship, which can be written as $[110]ZrO_2//[110]CeO_2$ and $(002)ZrO_2//(002)CeO_2$ as indexed in the fluorite type structure. Although some domain boundaries are visible within both ZrO_2 and CeO_2 layers, they are not strong enough to make any impact on XRD or SAEDP.

The strain generated as a result of lattice mismatch between ZrO_2 and CeO_2 was released by the interface dislocations, which are periodically distributed along the interface (HRTEM not shown here) Domain boundaries within the film for some occasions originate from the interface between the substrate and the first layer. This type of domain boundary consistently penetrates through the thickness of the film. There is no close correlation between the layer thickness and the density of dislocations. We noticed that the interface dislocations separation distance is larger than the value anticipated from the lattice mismatch. This was clearly due to the highly textured structure of the films and the internal dislocations that are formed in the bulk of the layered structures, both of which helped to relieve the strain caused by the lattice mismatch between the CeO₂ and ZrO₂ phases.

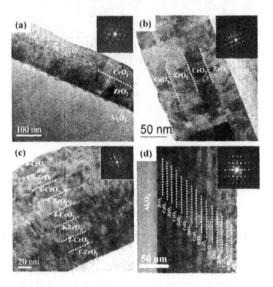

Fig. 8: TEM micrograph showing a cross sectional view of (a) a two-layer, (b) a four-layer, (c) an eight-layer and (d) a sixteen-layer Gd₂O₃-doped CeO₂ and ZrO₂ film grown on Al₂O₃(0001).

Oxygen ionic conductivity results for two-, four-, eight-, ten- and sixteen-layered GDC and GSZ films on Al₂O₃ (0001) substrates are displayed in Fig. 9. The oxygen ionic conductivity data from polycrystalline [28] and single crystal [29] 8 mole% YSZ are also shown for comparison. In general, these highly oriented films showed much higher conductivity compared to bulk polycrystalline 8 mole % YSZ. It is apparent from the sheet resistance measurements that increasing the number of interfaces (i.e., the number of discrete layers) in the structure facilitates ion transport and leads to an increase in the oxygen ionic conductivity at low temperatures with respect to a single layer of the same thickness. The ionic conductivities for single crystal 8 mole % YSZ and the two-, four-, eight-, ten- and sixteen-layered films, all at 650 K extracted from the Fig. 9(a) are shown in Fig. 9(b). At that temperature, increasing the number of layers resulted in higher oxygen ionic conductivity up to a thickness of 15 nm (for individual layers), beyond which conductivity decreases. The maximum value for the conductivity appears to be at least an order of magnitude higher than that from either polycrystalline gadolinia-doped bulk ceria or a single crystal yttria-stabilized zirconia thin film. However, when the thickness of individual layers was reduced below ~ 15 nm, the conductivity appears to decrease probably due to the strain effects associated with the thin films compared to relatively thicker films. An analogous increase in conductivity, with even greater magnitude, was observed in undoped fluoride superlattices [30], and was attributed to an increase in the density of carriers (fluorine vacancies and interstitials) due to space charge effects near the interfaces. However, this mechanism is unlikely to play a significant role in the oxide superlattices, since the Debye screening length is inversely proportional to the square root of the carrier density and is only approximately few Angstroms in this system; thus, the space charge region is very small. Here, enhanced ion conductance is expected to be a result of extended defects and lattice strain near the layer interfaces, which may increase the solubility of gadolinia, and hence the density of oxygen vacancies, in the ceria layers. Defects and strain relaxation may also increase the mobility of the vacancies.

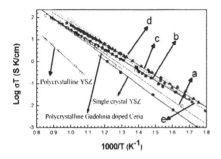 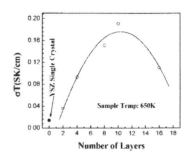

Fig. 9: left-a) Arrhenius plots of oxygen ionic conductivity of two- (a), four- (b), eight- (c), ten- (d) and sixteen-layer (e) gadolinia-doped ceria and zirconia thin films. The data from polycrystalline and single crystal YSZ are also displayed in this figure [Ref. 20]. right-b) Conductivities of single crystal YSZ [12], two-, four-, eight-, ten- and sixteen-layer films at 650 K [Ref. 20]

CONCLUSION:

We have used oxygen-plasma-assisted molecular beam epitaxy (OPA-MBE) to grow highly oriented $Ce_{1-x}Sm_xO_{2-\delta}$ films and gadolinia doped ceria/zirconia multi-layer films on single crystal c-Al_2O_3. All the films were characterized using several surface and bulk sensitive techniques. In samaria doped ceria films, the samarium concentration, x, was varied in the range 1-33 atom%. It was observed that dominant (111) orientation in $Ce_{1-x}Sm_xO_{2-\delta}$ films can be maintained up to about 10 samarium atom% concentration. Films higher than 10 atom% Sm concentration started to show polycrystalline features. The highest conductivity of 0.04 S.cm^{-1}, at 600 ^{0}C, was observed for films with ~ 5 atom% Sm concentration. In addition to the high quality epitaxial samaria doped ceria films, we have also prepared nanoscale multi-layer hetero-structures of doped ceria/zirconia to explore the conductivity mechanisms. Greatly improved ionic conductivity was obtained in these multi-layered structures compared to individual bulk electrolytes. Total conductivity is found to increase with the increasing number of layers. Although the exact mechanism associated with this enhanced conductivity remains unknown, strain enhancement of either dopant solubility or oxygen vacancy mobility could be responsible for the increase in the conductivity..

ACKNOWLEDGEMENT

The authors would like to thank S. Azad, M. E. Engelhard, D.E. McCready, A. El-Azab, J.S. Young, A. Dirkes and M. Mckinley for their help at various stages during the experiments. This research was supported in part by the Division of Chemical Sciences, Office of Basic Energy Sciences, U.S. Department of Energy and the Laboratory Directed Research and Development (LDRD) program. The experiments were performed in the Environmental Molecular Sciences Laboratory, a national scientific user facility located at Pacific Northwest National Laboratory (PNNL), and supported by the U.S. Department of Energy's Office of Biological and Environmental Research. PNNL is a multi-program national laboratory operated for the U.S. DOE by Battelle Memorial Institute under contract No. DE-AC06-76RLO 1830.

REFERENCES

1. M. S. Dresselhaus and I. L. Thomas, *Nature*, **414 (2001)** 332.
2. B. C. H. Steele and A. Heinzel, Nature, **414 (2001)** 345.
3. O.Yamamoto, Electrochemica Act., **45** (2000) 2423.
4. R. Doshi, V.L. Richards, J.D. Carter, X. Wang, and M. Krumpelt, J. Elec. Chem. Soc. **146** (1999) 1273.
5. H.L. Tuller, A.S. Nowick, J. Electrochem. Soc., **122** (1975) 255.
6. B. C. H. Steele, Solid State Ionics, **129** (2000) 95
7. M. Mogensen, N. M. Sammes and G. A. Tompsett, Solid State Ionics, **129** (2000) 63.
8. Z. Tianshu, P. Hing, H. Huang and J. Kilner, Sol. State Ionics, **148** (2002) 567.
9. G.Y. Meng, Q.X. Fu, S.W. Zha, C.R. Xia, X.Q. Liu, and D.K. Peng. Solid State Ionics, **148** (2002) 533.
10. H. Zhao and S. Feng, Chem. Mater., **11** (1999) 958.
11. E. Ruiz-Trejo, J.D. Sirman, Y.M. Baikov, and J.A. Kilner, Solid State Ionics, **113-115** (1998) 565.
12. T. Ohashi, S. Yamazaki, T. Tokunaga, Y. Arita, T. Matsui, T. Harami and K. Kobayashi, Solid State Ionics, **113-115** (1998) 559.
13. S.P.S. Badwal, F.T. Ciacchi and D. Milosevic, Solid State Ionics, **136-137** (2000) 91.
14. K. Nomura, Y. Mizutani, M. Kawai, Y. Nakamura and O. Yamamoto, Solid State Ionics, **132** (2000) 235.
15. T.I. Politova, and J.T.S. Irvine, Solid State Ionics, **168** (2004) 153.
16. M. Weller, F. Khelfaoui, M. Kilo, M.A. Taylor, C. Argirusis, and G. Borchardt, Solid State Ionics, **175** (2004) 329.
17. C. Haering, A. Roosen, H. Schichl, and M. Schnoller, Solid State Ionics, **176** (2005)261.
18. S. Sarat, N. Sammes, and A. Smirnova, J. Power Systems, **160** (2006) 892.
19. Z.Q. Yu, S. Kuchibhatla, L.V. Saraf, O.A. Marina, C.M. Wang, M.H. Engelhard, V. Shutthanandan, P. Nachimuthu, D.E. McCready, and S. Thevuthasan, Electrochemical and Solid State Letters, in press.
20. S. Azad , O.A. Marina, C.M. Wang, L.V. Saraf, V. Shutthanandan, D.E. McCready, A. El-Azab, J.E. Jaffe, M.H. Engelhard, C.H. Peden, and S. Thevuthasan, Appl. Phys. Lett. **86** (2005) 131906.
21. S. A. Chambers, T. T. Tran, T. A. Hileman, J. Mat. Res., **9** (1994) 2944.
22. L. J. van der Pauw, Philips Res. Rep, **13**, 1 (1958).
23.] X. Guo, R. Waser, Progress in Materials Science, **51**(2006)151.
24. H. Fujimori, M. Yashima, M. Kakihana, M. Yoshimura, J. Appl. Phys. **91**(10), 6493(2002).
25. J. M. G. Gutekunst, and M. Ruhle, Phil. Mag. A, **75** (1997) 1357.
26. J. M. G. Gutekunst, and M. Ruhle, Phil. Mag. A, **75** (1997) 1329.
27. S. C. J. J. R. Willis, and R. Bullough, Phil. Mag. A, **62** (1990) 115.
28. N.Q. Minh and T. Takahashi, *Science and Technology of Ceramic Fuel Cells*, Elsevier, Amsterdam, Lausanne (1995) 94.
29. S. Ikeda, O. Sakurai, K. Uematsu, N. Mizutani and M. Kato, J. Mater. Sci., **20** (1985) 4593.
30. N. Sata, K. Eberman, K. Ebert and J. Maier, Nature, **408** (2000) 946.

Interconnects

DEVELOPMENT OF NEW TYPE CURRENT COLLECTOR FOR SOLID OXIDE FUEL CELL

Tsuneji Kameda[1], Kentaro Matsunaga[1], Masato Yoshino[1], Takayuki Fukasawa[2], Norikazu Osada[2], Masahiko Yamada[3],Yoshiyasu Itoh[1]

1. Power and Industrial Systems R&D Center, Toshiba Corporation, Yokohama, Japan.
2. Corporate R&D Center, Toshiba Corporation, Kawasaki, Japan.
3. Power Systems Company, Toshiba Corporation, Tokyo, Japan.

ABSTRACT

A new type of current collector was developed and its bulk properties were evaluated. The effects on the electrical properties by applying it on the cathodic electrode side were also evaluated. The developed current collector was an electrically conductive porous ceramic consisted from an aggregate of the Ag thin layer covered spherical ZrO_2 particles. Its flexible green sheet was successfully formed. And it was well bonded on the electrodes by laminating and post sintering process, due to its properties of almost no shrinkage during sintering and of the approximately same thermal expansion coefficient to the typical cell constituent materials. It was recognized that this current collector application decreased ohmic loss by about 2/3 on the cathodic electrode side, without the degradations of the cathodic electrode activity or the loss effect from a gas diffusion disturbance. The material cost merit was also explained comparing the typical Pt and Ag mesh current collectors.

INTRODUCTION

Solid oxide fuel cell (SOFC) is attracting much interest due to the advantages such as the high energy conversion efficiency, the adaptability to many common hydrocarbon fuels without reforming, and the possibility to lower the cost by no use of precious metals, and so on. Additionally, the recent technical progresses on the study especially for lowering the operating temperature by employing the high ionic conductive electrolytes and also for validating the reliability of some stacks are drawing remarkable attention.

Although the cell itself is composed from an electrolyte and a pair of electrodes, current collectors, gas separators, manifolds and some others are also important constituent parts. All these parts are directly or indirectly relate to the cell and stack properties. In our previous evaluation for some fabricated cell samples, it was recognized that the ohmic loss at the current collector contact affected to the over all properties significantly. Figure 1 shows the typical result of analyzing the electrical loss dividing by certain contributing factors. It is clear that the improvement of the ohmic loss at the current collector contact is relatively important to enforce the cell electrical properties. In early study[1-4], SASAKI, et al.[1] investigated the geometrical effect of Pt-mesh current collector on the cathode polarization and ohmic loss as a function of current collector contact spacing, ranges from 10 to 0.5 mm. They reported the obvious decreasing tendency both of the cathode polarization and the ohmic loss by decreasing the current collector contact spacing. In this study, the new type of current collector with the contact spacing of 10μm order is developed and the effects of improving the electrical properties of cells are studied.

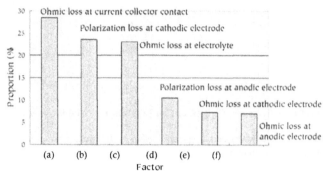

Figure 1. Typical pareto chart result analyzing the contributing factors on the electrical loss for a previous fabricated cell sample. (a); Ohmic loss at current collector contact. (b); Polarization loss at cathodic electrode. (c); Ohmic loss at electrolyte. (d); Polarization loss at anodic electrode. (e); Ohmic loss at cathodic electrode. (f); Ohmic loss at anodic electrode.

EXPERIMENT

The new type of current collector was a porous structured ZrO_2 ceramic, which was an aggregate of spherical ZrO_2 particles with electrically conductive Ag thin coating covered on the all surfaces of the particles. And the particles were connected each other by a small amount of Ag metal. The raw materials employed in this study are summarized in Table 1. The average diameter of the spherical ZrO_2 particles was about 15μm. After a surface treatment, thin film of Ag was coated by plating process. The thickness of the Ag coating was about 0.2μm. Then the 80wt% of Ag coated spherical ZrO_2 particles and 20wt% of Ag fine powder were mixed with some binders and a solvent. The green sheet with the thickness of 0.5mm was formed by doctor blade technique. The sheet was cut and laminated, and then sintered at 700°C to obtain the bulk samples for some properties evaluation. The sintered bulk samples were cut into 3mm x 4mm x 50mm bars and 3mm x 4mm x 15mm bars, and the electrical conductivity by four-terminal method and the thermal expansion coefficient by pushrod type method were measured. Also the heat resistant property of the coated Ag film on the spherical ZrO_2 particles was evaluated by SEM observation after several conditions of heat treatments.

The performance as the current collector on the cathodic electrode was evaluated. The electrolyte supported button cells of φ16mm diameter with and without developed current collector on the cathodic electrode were fabricated. The electrolyte was 8mol% Y_2O_3 stabilized ZrO_2 (8YSZ) of 300μm thickness. The anodic and the cathodic electrodes were NiO and Sm doped CeO_2 cermet (Ni-SDC), La-Sr-Co-Fe compound oxide and Sm doped CeO_2 mixture (LSCF-SDC), respectively. Some details of cell materials are summarized in Table 2. Both of the electrodes were fabricated by spray coating and sintering processes. The thicknesses were both about 30-50μm. After fabricating the complete configuration of the single button cell, the current collector of about 50μm thickness was laminated and sintered on the cathodic electrode side. The microstructure was observed by SEM for the cross cut surfaces. Cathodic polarization and ohmic loss were measured in air by the current-interrupt technique at 900°C.

Table 1. Specifications of raw materials used in this study

Spherical ZrO$_2$ particle
Average particle size : 15μm

Ag fine powder
Average particle size : 0.6-13.0μm
Specific surface area : 1.0-2.5 m^2/g

Table 2. Details of cell materials

Electrolyte
Material : 8mol% Y$_2$O$_3$ stabilized ZrO$_2$ (8YSZ)
Thickness : 300μm   Diameter : 16mm
Anodic electrode
Material : NiO and 20mol% Sm-CeO$_2$ cermet (Ni-SDC)
Thickness : 30-50μm   Diameter : 6mm
Cathodic electrode
Material : La$_{0.6}$Sr$_{0.4}$Co$_{0.2}$Fe$_{0.8}$O$_3$ and 20mol% Sm-CeO$_2$ (LSCF-SDC)
Thickness : 30-50μm   Diameter : 6mm

Thickness : 30-50μm   Diameter : 6mm

DEVELOPMENT AND PROPERTIES EVALUATION OF NEW TYPE CURRENT COLLECTOR

Figure 2 shows the photograph of the developed current collector green sheet. The thickness is about of 0.5mm. and it has good flexibility. After sintering. it is settled and show relatively high strength. It shows almost no shrinkage during sintering. So this developed current collector is thought to be able to laminate on the completely configured cell, avoiding the bothersome co-firing process. Figure 3 shows the microstructure of this sintered sheet. The structure composed of Ag coated spherical ZrO$_2$ particles and binding Ag fine grains is observed and it is recognized that the networks of the open pores are formed in preferable way. Figure 4 shows the results of heat resistant test for Ag coated spherical ZrO$_2$ particles. The particles were heat treated in the air at the temperature from 700 to 900°C for 1 hour. The coated Ag films are kept the original covering shape on the ZrO$_2$ particles up to 900°C. The long term test will be planned.

Figure 2. Developed current collector green sheet.

Figure 3. Microstructure of the sintered current collector.

Figure 4. Results of heat resistant test for Ag coated spherical ZrO$_2$ particles. After heat treatments for 1 hour in air at (a);700°C, (b);900°C.

Figure 5 shows the electrical resistance of the sintered current collector comparing some other electrical conductive materials. The electrical resistance of the developed current collector is low enough compared with typical cathodic electrode materials. Figure 6 shows the thermal expansion coefficient of the sintered current collector. The thermal expansion coefficient is almost same to that of 8YSZ. This is thought to because the current collector is consisted from a framework structure of the spherical 8YSZ particles. So it is expected to have good coherence with the cell constituent materials.

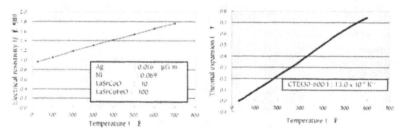

Figure 5. Electrical resistance of the sintered current collector compared with some other electrical conductive materials.

Figure 6. Themal expansion coefficient of the sintered current collector.

CELL PROPERTIES WITH DEVELOPED CURRENT COLLECTOR

Figure 7 shows the microstructure of the cross section of the button cell, which is laminated the current collector on the cathodic electrode. The porous structure in the current collector and a good bonding to the cathodic electrode are recognized. The ohmic loss and the polarization properties for the cathodic electrode side of the button cells with and without the developed current collector are shown in Figures 8 and 9. They were measured at 900°C. It is recognized that the ohmic loss properties of the developed current collector applied cells show about 2/3 lower than that of the usual cells without the current collector, which means using Pt-mesh for contact. On the other hand, the polarization properties show almost no difference for the cells with and without the developed current collector, so the degradations of the cathodic electrode activity and of the effect from gas diffusion loss are thought to be negligible small.

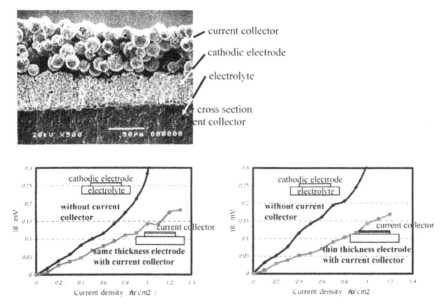

Figure 8. Ohmic loss properties for the cathodic electrode with and without the developed current collector at 900°C.

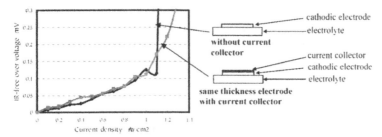

Figure 9. Polarization properties for the cathodic electrode with and without the developed current collector at 900°C.

Figure 10 shows the estimated comparison of the material costs for Pt-mesh, Ag-mesh and the developed current collector. The material costs are based on the values of 50 dollar/g for Pt, 0.5 dollar/g for Ag, and 0.1 dollar/g for ZrO_2. It is thought to be cost effective to apply the developed current collector, as well as its benefits of good electrical conductivity.

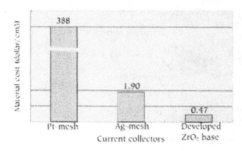

Figure 10. Estimated comparison of the material costs for some typical and the developed current collectors.

CONCLUSION

The new type of current collector, which was mainly consisted from an aggregate of the Ag thin layer covered spherical ZrO_2 particles, was developed and its properties were evaluated. The evaluation tests for bulk samples of the current collector reveal that its electrical resistant is low enough comparing typical cathodic materials and the thermal expansion coefficient is near to the typical cell constituent materials. Also this current collector shows almost no shrinkage during sintering. These properties are good for laminating it on the cell by post sintering process. Even for the button cell size test, it is confirmed that the current collector application decrease the ohmic loss by about 2/3 on the cathodic electrode side, without certain other obstacles. The material cost merit is also expected for this current collector.

REFERENCES

[1]K. Sasaki, L. J. Gauckler, Microstructure-Property Relations of Solid Oxide Fuel Cells, Microstructural Design of Cathodes and Current Collectors, Denki Kagaku, 64, [6] 654-661 (1996)
[2]M. Guillodo, P. Vernoux, J. Fouletier, Electrochemical properties of Ni-YSZ cermet in solid oxide fuel cells, Solid State Ionics, 127, 99-107 (2000)
[3]S.P. Jiang, J.G. Love, L. Apateanu, Effect of contact between electrode and current collector on the performance of solid oxide fuel cells, Solid State Ionics, 160, 15-26 (2003)
[4]Y. Itagaki, F. Matsubara, M. Asamoto, H. Yamaura, H. Yahiro, Y. Sadaoka, Electrophoretically Coated Wire Meshes as Current Collectors for Solid Oxide Fuel Cell, ECS Transactions, 7, (1) 1319-1325 (2007)

ELECTRICAL CONDUCTIVITY AND OXIDATION STUDIES OF CERAMIC-INTERMETALLIC MATERIALS FOR SOFC INTERCONNECT APPLICATION

Yukun Pang, Hua Xie and Rasit Koc

Department of Mechanical Engineering and Energy Processes, Southern Illinois University at Carbondale; Carbondale, IL, 62901, USA

ABSTRACT

Significant progress in lowering the operating temperature of SOFCs demands for new materials systems with further improvements that are not attainable in oxide-based and metallic alloys. The present work deals with the development of a new material system based on $TiC-Ni_3Al-TiN$. The electrical conductivity and oxidation resistance of $TiC-Ni_3Al$ and $TiN-Ni_3Al$ cermets were investigated to evaluate the feasibility for SOFC interconnect application. TiC and TiN cermets with 30, 50 and 70 wt% of Ni_3Al were fabricated using powder processing steps, and were then subjected to electrical conductivity measurements in air at 800 °C for 100 hours. Although all samples tested were slightly oxidized, no degradation has been observed and the oxide scales strongly adhered to the samples and shielded them from further oxidation. High electrical conductivity values of the order of 10^3 S/cm have been recorded in the range of the measurements. With reference to the mixture rules for electrical conductivity, SEM and EDS results were interpreted to discuss the impact of the oxide scales on the electrical conductivities.

INTRODUCTION

Solid oxide fuel cells (SOFCs) have been the subject of intensive research and development around the world over the past several decades. They are considered as one of the most efficient energy conversion devices with zero emissions. Under typical SOFC operating conditions, a single cell produces less than 1 Volt [1]. To obtain high voltage and power from SOFCs, it is necessary to stack many cells together and this can be done in a number of ways using interconnect materials. Regardless of designs, there are three primary functions of interconnects: (1) to provide electrical connection between the cathode of one single cell to the anode of the adjacent single cell; (2) to isolate the fuel and air supplies for each cell. One of the obstacles for the commercialization of SOFCs is to develop interconnect materials with adequate electrical conductivity and long-term stability under operating conditions.

In fact, for operation temperature above 800°C, doped lanthanum chromite ($LaCrO_3$) is the most common used material for SOFC as it has relatively high electrical conductivity at high temperatures and good stability in both oxidizing and reducing conditions [2]. However, the major disadvantages of this material include high fabrication and material costs. The reduction of the cell operating temperature down to below 800°C makes the use of metallic materials for interconnect feasible and attractive. The advantages of metallic interconnect over ceramic interconnects include better electrical and thermal conductivity, lower material and fabrication costs, and easier processing. The two main disadvantages of metallic interconnects are: (1) the formation of volatile chromia on the surface of chromium-based metallic interconnect can lead to severe degradation of the electrochemical performance of SOFC by poisoning the cathode; (2) the formation of oxide scale leading to significant ohmic losses [1,3,4].

Significant progress in lowering the operating temperature of SOFCs demands for new materials systems with further improvements that are not attainable in oxide-based and metallic alloys. The intention of using compositions of TiC-Ni₃Al and TiN-Ni₃Al is to combine the advantages of ceramics and intermetallics. A summary of the relative properties of TiC, TiN and Ni₃Al is shown in Table I [5-13]. As shown in the table, it is clear that TiC and TiN have high thermal stability and high electrical conductivity at room temperature.Ni_3Al, when used as a binder phase to fabricate dense, is reported to be able to improve the oxidation resistance of the composites [13]. Further more, when Ni_3Al is oxidized into NiO and Al_2O_3, the electrical conductivity of nickel oxides is very high compared with that of chromia at elevated temperatures [5], and the Al_2O_3 content is expected to be trivial so that its impact on the electrical conductivity could be omitted. Thermal expansion coefficient is expected to be compatible with other cell components or it can be easily tailored by adjusting the content proportion.

The electrical conductivity of these cermets under oxidizing and reducing conditions at high temperature as well as the correlative oxidation behavior have not been reported in literature. In this study, the electrical conductivity and oxidation behavior of TiC- Ni_3Al and TiN- Ni_3Al cermet at 800°C in air for a period of 100 hours were initially investigated in order to assess the feasibility for SOFC interconnect application.

Table I. Properties of TiC, TiN and Ni

Properties	TiN	TiC	Ni₃Al	YSZ Electrolyte
Density, g/cm³	5.39	4.92	7.5	
Melting Point, °C	2930	3065	1390	
Thermal conductivity, at 1073 K, W/m•K	26.7	41.8	36.1	
Thermal Expansion Coeff., 1/°C x 10⁻⁶	7.4	7.4	~15	~10.5
Electrical Conductivity (RT), S/cm x 10⁴	1-4	3.3	2.6	Ionic

EXPERIMENTAL PROCEDURE

Nano-sized TiC powders with high purity and high surface area were made from 29.5 wt% carbon coated TiO_2 precursors synthesized using a patented process developed by R. Koc and G. Glatzmaier [11,12]. TiN powders (Grade C, H.C. Starck) with average particle size of 0.8-1.2 μm were purchased. TiN and TiC powders were mixed with 30, 50, 70 weight percent of Ni₃Al powders (-100 mesh, Alfa Aesar), respectively, using wet milling in tungsten carbide (WC) container with WC balls and alcohol for 2 hours. Powder mixtures were sent to Greenleaf Corp. where they were sintered by hot pressing. TiC- Ni₃Al samples were sintered in the range of 1350 to 1490 °C for 3600 seconds in vacuum (~50 mTorr) with sintering pressure of 4500 psi. TiN- Ni₃Al samples did not quite densify when sintered in the temperature range from 1200 to 1300 °C for 3600 seconds in vacuum (~50 mTorr) with sintering pressure of 4500 psi, and were re-sintered at 1400 °C and 3000 psi for 1800 seconds in vacuum (~50 mTorr). Hot pressed samples were furnace cooled and were ground or polished to eliminate the graphite residue on surfaces, and then were subjected to density measurements and X-ray diffraction (XRD). Both TiC- Ni_3Al and TiN- Ni_3Al mixtures achieved over 80% of theoretical density. Small bars (about $2.5 \times 2.5 \times 12.5$ mm^3, labeled as TiN-30/50/70 Ni_3Al for TiN-30/50/70 wt% Ni_3Al and TiC-30/50/70 Ni_3Al for TiC-30/50/70 wt% Ni_3Al, respectively) were cut from the ground or polished samples. Weight and dimension of the bars were recorded.

The bar samples were, one at a time, subject to four-wire electrical conductivity measurements utilizing a Linear Research Inc LR-700 AC Resistance Bridge operating at a frequency of 16 Hz. The sample was mounted in the resistance bridge with no electrode paste. Electrode paste was not applied because (a) samples have high conductivities at room temperature, (b) the cross section area of the bars are close to that of the resistance bridge tip, and (c) the electrode paste might interfere the oxidation behavior of the materials. Samples were heated from room temperature (24°C) to 800°C at a speed of 5 to 7°C per minute using a Carbolite Model CTF 17/75/300 Tube Furnace, hold at 800°C for 100 hours, and then furnace cooled. An S-type thermocouple was placed close to the sample to read temperature. The electrical resistance was recorded during the period of 100 hours at 800°C. The data of electrical conductivity were then calculated and plotted using Microsoft Excel. Each test lasted for totally 120 hours, including 15 hours for heating from room temperature to 800 °C, 100 hours for holding at 800 °C, and 5 hours for furnace cooling. Weight gains of the tested samples were measured. XRD analyses were performed on the surfaces of oxidized samples to investigate the composition of the oxidation scale. Scanning electron micrographs were taken on the polished cross-section of the oxidized samples to study the microstructure and to measure the thickness of the oxidation scale. Energy dispersive x-ray spectroscopy (EDS) analyses were utilized to study the composition of the oxidation scale. The samples were mounted in epoxy resin for polishing and remained so during SEM and EDS analyses.

RESULTS AND DISCUSSION

Oxidation Behavior

Scales were found on all samples after electrical conductivity measurements in air at 800°C. XRD spectra of the surfaces of TiN- Ni_3Al and TiC- Ni_3Al samples showed similar XRD patterns, respectively. Therefore, spectra of 50 wt% Ni_3Al samples were used to represent the two series, as shown in Fig. 1. Peaks of TiO_2, NiO, and $NiTiO_3$ have been identified in spectra of TiN- Ni_3Al, whereas only TiO_2 and NiO in those of TiC- Ni_3Al. This indicates that Ni_3Al and TiN or TiC at the surface were oxidized during the measurements. According to the study of M. Haerig and S. Hofmann [14], aluminum oxide could probably form during the oxidation of Ni_3Al. In this study, however, no peak of Al_2O_3 has been identified in XRD analyses, as shown in Fig. 1. A possible explanation for the absence of Al_2O_3 peaks is that the amount of Al_2O_3 in the scale is too small to be detected by the XRD equipment.

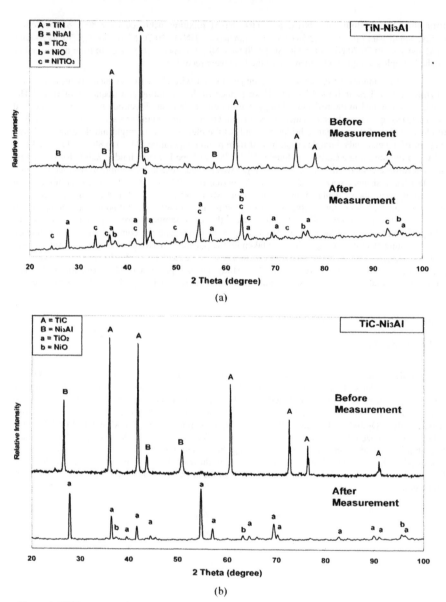

Figure 1: XRD patterns of (a) TiN- Ni₃Al and (b) TiC- Ni₃Al samples before and after electrical conductivity measurements.

Table II. Weight Gain Of Samples After Electrical Conductivity Tests

Composition	TiC			TiN		
	30Ni$_3$Al	50Ni$_3$Al	70Ni$_3$Al	30Ni$_3$Al	50Ni$_3$Al	70Ni$_3$Al
Percentage weight gain	1.80%	2.40%	2.90%	3.80%	0.82%	0.79%
Oxide scale thickness (μm)	55-60	55-60	55-60	10-15	7-8	5

Fig. 2 and 3 are the Backscattered electron micrographs of polished cross section of oxidized samples where "I" is the epoxy resin, the gray-white band "II" the oxide scale, and "III" the cermet substrate,. It can be seen in Fig. 2 that there are areas with different grayscale in the oxide scale of TiN- Ni$_3$Al samples, indicating the formation of different oxides of Ti and Ni. Given that the atom number of Ni (28) is greater than that of Ti (22), the brighter areas in the scales are believed to contain higher Ni-containing oxide or Ni content, which is in accordance with the EDS analyses results. Base on EDS analyses, different features in the backscattered electrons microphotographs of polished cross section of TiN- Ni$_3$Al samples were labeled as follows: the brighter area "a" is oxides with relatively high Ni content, the grey area "b" is the mainly TiO$_2$, the bright white area "c" is intermetallic Ni$_3$Al phase, and the gray area "d" is the TiN grain.

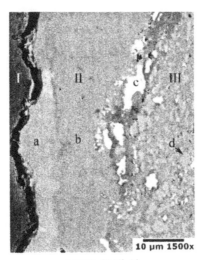

(a) TiN-30 Ni$_3$Al

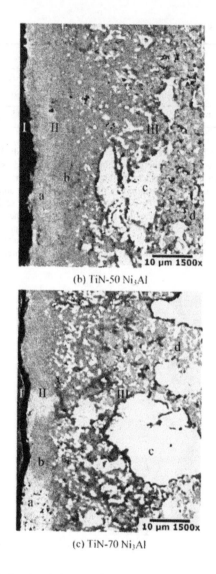

(b) TiN-50 Ni₃Al

(c) TiN-70 Ni₃Al

Figure 2: Backscattered electron micrographs of polished cross section of oxidized TiN- Ni₃Al samples, I) Epoxy Resin, II) Oxide scale, III) Cermet.

As shown in Fig. 3, the oxide scale of TiC-Ni$_3$Al samples consist of two layers and some particles showing the traces of diffusion. Further EDS analyses have revealed that: TiO$_2$ is the main ingredient of the oxide scales; layer L, which can be easily observed at the scale margins of all three TiC-Ni$_3$Al samples, has higher Al and much higher Ni content than the rest of the oxide scale; the particles P are traces of unoxidized cermet granules diffusing toward the scale surface. Compared to TiN-Ni$_3$Al samples, the content of such Ni-containing oxide is much lower in the oxide scale of TiC-Ni$_3$Al samples.

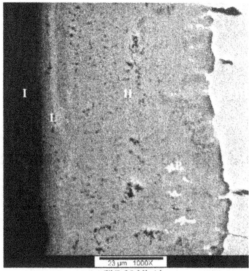

(a) TiC-30 Ni$_3$Al

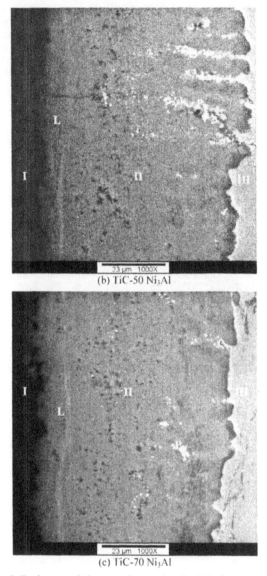

(b) TiC-50 Ni₃Al

(c) TiC-70 Ni₃Al

Figure 3: Backscattered electron micrographs of polished cross section of oxidized TiC-Ni₃Al samples, I) Epoxy Resin, II) Oxide scale, III) Cermet.

The oxide scales of all samples were strongly adherent to the cermet substrates. No crack or spalling of the scales occurred during heating circle or even cutting and polishing process. All samples gained a small quantity of weight as shown in Table II. The moderate weight gain indicates that the oxide scale, once formed, acted as a protective layer shielding the specimen from further oxidation. Another interesting point is that TiC samples with higher Ni_3Al content gained more weight while it was the opposite case for TiN samples. As for TiC samples whose oxide scales are of similar thickness, more weight gain could be a result of more Ni or NiO content in the scale. TiN samples with more Ni_3Al, on the other hand, had apparently thinner oxide scales resulting in less weight gain. It is also worth mentioning that the oxidation of TiN-Ni_3Al samples started at a temperature 100°C higher than TiC-Ni_3Al ones. Generally, TiN-Ni_3Al showed better oxidation resistance than TiC-Ni_3Al did.

Electrical Conductivity

Electrical resistance of the sample was recorded during the measurement and then calculated into electrical conductivity, σ in Siemens/cm (S/cm), using Equation 1:

$$\sigma = \frac{1}{\rho} = \frac{l}{RA} \tag{1}$$

where ρ is the electrical resistivity, R the resistance (in ohms) directly read from the resistance bridge, l the length of the sample (in cm), and A the cross-section area of the sample (in cm^2). The electrical resistivity of an oxidized sample combines resistivities of the cermet substrate and the oxide scale according to the electrical conductivity mixture rules. In comparison, however, the resistivity of the cermet is so small that the resistivity of the oxide scale dominates the electrical resistivity and thus the conductivity of the oxidized sample.

The electrical conductivities were calculated and plotted as σ versus time in hours, as shown in Fig. 4. Apparently, samples with higher intermetallics content exhibited better electrical conductivity. It is believed to be a result of more NiO content in the oxide scale of cermets with higher weight percent of Ni_3Al, as the electrical conductivity of NiO is much higher than the other oxides in the scale, TiO_2, $NiTiO_3$ and Al_2O_3 [1, 15]. It can also be seen in Fig. 4 that the electrical conductivities of TiN-70Ni_3Al and TiN-30Ni_3Al increase slowly with time during the 100 hours tests. The improvement of the electrical conductivity is believed to be a result of the increase of the volume fraction of NiO in the oxide scale outbalancing the thickening of the scale. The electrical conductivity of those samples will decrease if the negative effective of the later process outbalances the positive effect of the former one during an extended testing time.

Further more, the electrical conductivities of TiC-Ni_3Al were in a trend of decreasing value, which is believed to be the result of the growth of the oxidation scale. Nonetheless, the conductivities of all samples remained at a very high level at the end of the 100 hours tests, compared to that of the cathode and other types of interconnects, as shown in Table III. The electrical conductivities of TiN-Ni_3Al, on the other hand, were more stable with no obvious trend of decreasing. In general, TiN- Ni_3Al samples exhibited much better electrical conductivities than TiC- Ni_3Al ones, which are believed to be due to TiN- Ni_3Al samples' better oxidation resistance and thus much thinner oxide scales as well as the higher NiO content in the oxide scales.

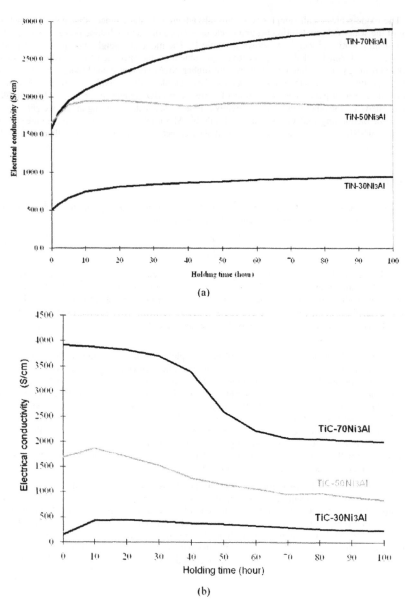

Figure 4: Electrical conductivity of (a) TiN-Ni₃Al and (b) TiC-Ni₃Al samples as a function of time (from 0 to 100 hours) at 800°C in air.

Table III. Comparison of the electrical conductivity of the cathode and different types of interconnect [4.16]

Composition	Electrical conductivity in oxidizing atmosphere (S/cm)	Temperature (°C)
Sr-LaMnO$_3$ (Cathode)	150	1000
LaCr$_{0.95}$Mg$_{0.05}$O$_3$	3.2	1000
Stainless steel	22-44 (Oxide scale dominates the electrical resistance)	800
TiN-70wt% Ni$_3$Al	2900	800
TiC-70wt% Ni$_3$Al	2000	800

Since no electrode paste was used in order not to interfere the oxidation behavior of the samples, the experimental contact area A' between the sample and the round tip of the resistance bridge was smaller than the cross-section area A used in the electrical conductivity calculation, as shown in Fig. 5. Thus, the real electrical conductivity of the sample should be even greater than the calculated one in this study. Electrical conductivity measurements with electrode paste on surfaces of oxidized samples are needed to further investigate the impact of the oxide scale composition on the area specific resistance (ASR).

Figure 5. Schematic drawing of the contact between the sample and the resistance bridge tip.

CONCLUSION

TiN-Ni$_3$Al and TiC-Ni$_3$Al cermet materials have shown great potential for SOFC interconnect applications and are believed to be more suitable than metallic alloys in terms of electrical conductivity and oxidation behavior in air at 800 °C. The oxide scales, once formed, were strongly adhesive to the cermet substrate and acted as a protective layer in oxidation conditions protecting the samples from further oxidation. Consequently, these materials were stable in oxidizing conditions at 800°C with no degradation. The electrical conductivity values of 2900 S/cm for TiN-70wt% Ni$_3$Al and 2000 S/cm for TiC-30wt% Ni$_3$Al in oxidizing condition at 800 °C was recorded. The content of NiO or Ni in oxidation scale was believed to have a significant and positive impact on the electrical conductivity of the tested samples. The conductivity of TiC-Ni₃Al remained at a very high level after a slow reduction during the measurement. The reduction in electrical conductivity was believed to be the result of the growth of oxide scales on the surface. The electrical conductivities of TiN- Ni₃Al, on the other hand, were more stable with no obvious trend of decreasing. Generally, TiN samples exhibited better electrical conductivity due to thinner oxide scale and higher NiO content in the scale.

ACKNOWLEDGEMENTS

The authors would like to thank Dr. Jason Goldsmith at Greenleaf Corp. for his help in sample preparation. Partial support from MTC is also acknowledged.

REFERENCES

[1] W.Z Zhu and S.C. Deevi, Development of interconnect materials for solid oxide fuel cells, *Materials Science and Engineering*, A 348, 227-234 (2003).

[2] R. Koc, and H.U. Anderson. Electrical Conductivity and Seebeck Coefficient of (La, Ca) (Cr, Co)O₃, *Journal of Material Science*, 27 (20), 5477-82 (1992).

[3] L. Chen et al., Clad metals, roll bonding and their applications for SOFC interconnects, *Journal of Power Sources*, 152, 40-45 (2005).

[4] J. W. Fergus, Metallic Interconnects for Solid Oxide Fuel Cells, *Materials Science and Engineering*. A 397, 271-283 (2005).

[5] W.Z. Zhu and S.C. Deevi, Development of Interconnect Materials for Solid Oxide Fuel Cells, *Materials Science and Engineering*, A 348, 227-243 (2003).

[6] Y.S. Touloukian et al., Thermophysical Properties of Matter. vol. 2 (IFI/Plenum, New York), 594,668 (1970)

[7] Y.S. Touloukian et al., Thermophysical Properties of Matter. vol. 1 (IFI/Plenum, New York), 237 (1970).

[8] R. Koc and J.S. Folmer, Synthesis of Submicrometer Titanium Carbide Powders, *J. Am. Ceram. Soc.*, 80), 952-956 (1997.

[9] A.Tarniowy, R. Mania, and M. Rekas, The Effect of Thermal Treatment on the Structure, Optical and Electrical Properties of Amorphous Titanium Nitride Thin Films, *Thin Solid Films*, 311, 93–100 (1997).

[10] J. Gunnars and U. Wiklund, Determination of Growth-Induced Strain and Thermo-Elastic Properties of Coatings By Curvature Measurements. *Materials Science and Engineering*, A 336, 7–21 (2002).

[11] R. Koc and G. Glatzmaier, US Patent No: 5,417,952.

[12] R. Koc, C. Meng, and G. A. Swift, Sintering Properties of Submicron TiC Powders from Carbon Coated Titania Precursor, *Journal of Materials Science*, 35, 3131-41 (2000).

[13] Becher. P. F. and Plucknett, K. P., Properties of Ni₃Al-bonded Titanium Carbide Ceramics, *Journal of the European Ceramic Society*, **18** (4), 395-400 (1998).

[14] M. Haerig and S. Hofmann, Mechanisms of Ni₃Al Oxidation between 500°C and 700°C, *Applied Surface Science*, 125, 99-114 (1998).

[15] R.S. Singh, T.H. Ansari, R.A. Singh and B.M. Wanklyn, Electrical Conduction in NiTiO₃ single crystals, *Materials Chemistry and Physics*, 40, 173-177 (1995).

[16] R. Koc and H.U. Anderson, Investigation of Strontium-doped La(Cr,Mn)O₃ for Solid Oxide Fuel Cells, *Journal of Materials Science*, 27, 5837-43 (1992).

Seals

IMPROVEMENT IN INTERFACE RESISTANCE OF CONDUCTIVE GAS-TIGHT SEALING MATERIALS FOR STACKING MICRO-SOFC

Seiichi Suda, Koichi Kawahara, Kaori Jono and Masahiko Matsumiya
Japan Fine Ceramics Center
Nagoya, Aichi, Japan

ABSTRACT

Micro-sized SOFCs are required small and correctly positioned interconnects, which work as gas-tight sealing and simultaneous electrical connection between cathodes and anodes in order to minimize electrochemical stacking losses and fabricate compact SOFC systems. We have developed highly conductive gas-tight seals that melt at lower than 850°C in air and maintain electrical conductive connection between both electrodes at 650°C in air. The addition of Si into Ag-Ge alloy much improved oxidation resistance and Ag-Si-Ge alloy resulted in highly conductive gas-tight seals after melting at 850°C when the alloy was fused on dense ceramics or metal manifolds. Ag-Si-Ge alloy had eutectic structure, which was composed two regions; one region contained silver and the other was composed of Si-Ge alloy. Ag region in the alloy would play a role in high electrical conduction, whereas Si-Ge region would improve oxidation resistance. Adsorbed or residual oxygen in the alloy would be trapped in the Si-Ge region, and this oxygen getting region would prevent highly conductive Ag region from oxidation or degradation of electrical properties. However, the fusion of Ag-Si-Ge alloy on LSCF porous cathode materials partly led to form thin glassy interfacial phase that showed relatively high resistance. Ag-containing paste was then developed to decrease the interfacial resistance. The coating with the Ag paste on Ag-Si alloys much reduced the interfacial resistance. Ag-Si alloy coating with Ag paste would be suitable for the highly conductive gas-tight sealing devices applicable for stacking micro-sized SOFCs.

INTRODUCTION

Downsizing of SOFC cells, stacks and systems is a great attractive for the purpose of improvement of cost performance, decrease in occupied areas and application as portable electric generators. High-efficient SOFC stacks are surely constructed with high-performance single cells and interfacial devices. The design of interfacial devices is surely altered by stacking design. In our project[1], the devices are required to work as gas-tight seals and electrical connections between cathodes and anodes[2]. Stacking with high-performance cells results in highly efficient SOFC system unless stacking loss is negligible small. However, the losses are rather large, and development of the interfacial devices is thus indispensable to obtain highly efficient small-sized SOFC systems. Development of these devices contains some challenges. The devices, which work as interconnects and gas-tight seals, are then required to be relatively thin and fixed with electrodes by fusion. The devices are also desired to possess low interfacial resistance between the devices and electrodes as well as high conductivity to decrease in stacking losses. Permitted space for the interfacial devices is small to construct compact SOFCs. Small SOFC system would be used many these interfacial devices, but each contact area between the device and electrodes is limited small.

We have developed highly conductive gas-tight sealing devices for stacking micro-sized SOFC. Our main target properties of the developing sealing devices were set as follows. The devices possess (1) a fused temperature of less than 850°C, (2) high oxidation resistance at 650°C in air, (3) conductivity as high as commonly used metals have, and (4) low interfacial resistance. After examining these aimed properties and some conductive materials, we chose some Ag alloys and discussed whether use of these alloys can match our demands. Ag metal showed relatively high oxidation resistance at 650°C in air, but the melting temperature of Ag is above 950°C and the temperature was too high to satisfy our target fusion temperature. Ag-Ge alloy showed lower fusion temperature than 700°C but poor oxidation resistance. The addition of Si into Ag-Ge alloy much improved oxidation resistance. Ag-Si-Ge alloy was melted at 850°C by controlling Ag/Si/Ge ratios and this alloy resulted in conductive gas-tight sealing when fused on dense ceramics or metal manifolds[3]. Ag-Si-Ge alloy showed eutectic structure, which was composed of two regions; one region contained silver and the other was Si-Ge alloy. Ag region in the alloy would fulfill a role of high electrical conduction, whereas Si-Ge region would improve oxidation resistance. Adsorbed or residual oxygen in the alloys would be trapped in the Si-Ge region, and this oxygen getting region would prevent highly conductive Ag region from oxidation or degradation of electrical properties.

However, when Ag-Si-Ge alloy was fused between LSCF porous ceramics, the fused samples showed relatively high resistance. Actually the area specific resistance (ASR) of fused Ag-Si-Ge alloy on Pt plate was measured to be lower than 4.0 mΩ cm^2, but that on porous LSCF ceramics was larger than 100 mΩ cm^2. We thus investigated why the alloy showed high interfacial resistance on porous LSCF and some coating on the alloys was investigated to decrease the interfacial resistance. Ag-containing paste was developed for the coatings on Ag-Si-Ge and Ag-Si alloys and effects of the paste coating on interfacial resistance was discussed to obtain interfacial devices suitable for stacking of micro-SOFCs.

EXPERIMENTAL PROCEDURE

Various Ag alloys were prepared by melting Ag, Si and Ge metals at 1500°C for 2 h in 4%H$_2$-Ar atmosphere, and the alloy foils were obtained by metal-rolling at room temperature or 360°C. The compositions of these alloys were set to be Ag-2.99mass%Si-7mass%Ge (Ag-Si-Ge), Ag-2.99mass%Si (Ag-Si), and Ag metal. Porous La$_{0.6}$Sr$_{0.4}$Co$_{0.2}$Fe$_{0.8}$O$_3$ (LSCF) ceramics were synthesized by sintering at 1330°C for 2 h in air. Alloy foils were placed between porous LSCF ceramics, and the sandwiched test pieces were fused at 800°C or 850°C for 2 h in air. Microstructures of the alloys after the fusion were observed with an optical microscope and a scanning electron microscope (SEM).

Ag-containing paste was synthesized by mixing Ag fine powder (average particle size: 2 μm), glass precursor particles and some organic binders. The volume ratio of Ag powder to glass precursor particles was adjusted to be Ag/glass = 6/4 with the view of connecting Ag powder in the fused paste on the basis of percolation theory. The glass precursor particles were mixed with two kinds of particles. Amorphous Na$_2$O-SiO$_2$ spherical particles (NS particles) were prepared by a sol-gel process, which is allowed to ion-exchange between sodium ion and proton derived from OH groups that were formed by hydrolysis of tetraethyl orthosilicate (TEOS) just before the condensations[4]. TEOS, hydroxyl propylcellulose and concentrated NaOH solution was mixed with ethanol and stirred vigorously at 50°C in air. Stirring for 1 h resulted in suspension dispersed with NS particles. Na/Si ratio of NS particles was estimated to be 0.63 from the results by energy-dispersive X-ray spectroscopy (EDS). The other precursor particles were sodium-free amorphous SiO$_2$ spherical particles (SI particles). SI particles were synthesized by a similar method to the preparation of NS particles. The preparation of SI particles was used NH$_3$ solution instead of the concentrated NaOH solutions. NS particles were impregnated with water at room temperature for 24 h and calcined at 300°C for 12 h in air. The calcined NS particles and SI particles was mixed at the mass ratios of NS/SI = 100/0, 90/10 and 80/20. Ag-containing pastes using NS and SI particles at the ratios NS/SI=100/0, 90/10, 80/20 were abbreviated to 100/0 paste, 90/10 paste and 80/20 paste, respectively.

Whether some reaction products were formed at the interface between Ag-paste and LSCF was investigated using XRD. Ag-pastes with various NS/SI ratios and LSCF powder were mixed using ethanol and the mixture was fired at 800°C for 2 h in air. XRD profiles were measured for the fused samples after being grounded with an alumina mortar.

ASR was then estimated for some test pieces. Ag-Si or Ag-Si-Ge alloy foil was placed between porous LSCF ceramics with or without the Ag paste, and the sandwiched samples were fused at 800-850°C for 2 h in air. Resistance for the samples was measured by dc 4-probe method, and ASR was estimated by the dependence of resistance on distance between inner two electrodes (Fig. 1). Current density was passed in the range of -1.0 to 1.0 A between outer two electrodes and voltages between inner two electrodes were measured at 400-650°C. Therefore the ASR value estimated in this study contains resistance Ag-Si alloy and Ag paste and ASR of LSCF/(Ag paste) and (Ag paste)/(Ag-Si alloy).

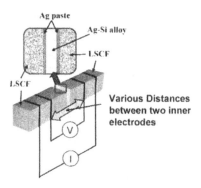

Fig. 1. Schematic illustration of measurement of ASR.

RESULTS AND DISCUSSION

Ag-Si-Ge alloy had relatively low fusion temperature and high oxidation resistance and was a candidate for high conductive interfacial devices. Ag-Si-Ge alloy was then placed between LSCF porous ceramics and fused at 850°C in air. Fig. 2 shows SEM image of fused Ag-Si-Ge alloy. The

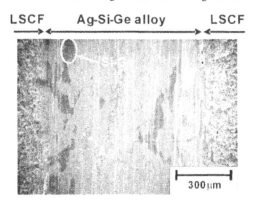

Fig. 2 SEM image of Ag-Si-Ge alloy fused between porous LSCF ceramics

fusion at 850°C resulted in tight connection between porous LSCF and the alloy, and would exhibit gas-sealing properties. EDS profile showed grey region in the alloy was composed of Ag and black region was Si-Ge alloy. Fig. 2 exhibited that the Si-Ge region was localized in the vicinity of LSCF. The Si-Ge region would slightly contain some oxides such as glassy phases, which show rather low conductivity. This localized Si-Ge region would bring about high ASR.

When Ag-Si-Ge alloy was fused between porous LSCF ceramics, the highly electrical resistive region was localized at the interface between the alloy and LSCF. Coating on the surface of the alloy was then investigated to avoid direct connection between the alloy and LSCF during the fusion. Ag pastes that were composed of fine Ag particles, glass precursor particles and some organic binders were investigated to decrease ASR on porous LSCF. The glass precursor was mixed with NS particles and SI particles. Ag pastes with various NS/SI ratios were prepared to investigate whether the glass precursor was applicable as coating materials. When flexible sealing sheet using NS particles and SI particles were previously developed to obtain highly resistive gas-tight seals, we found that $NaLa_9(SiO_4)_6O_2$ (NLSO) was slightly formed by the fusion with NS particles and LSCF powder at 800°C[5]. NLSO showed much lower electrical conductivity than LSCF, and this phase was increased with increasing in fusion time. Formation of the NLSO phase was also much depressed by the addition of SI particles. The mechanism of decrease in NLSO phase by the addition of SI particles is still vague, but the formation of NSLO phase would be much influenced by the diffusion of sodium during the fusion as shown in Fig. 3. Coating of Ag-containing paste was expected to lead the electrical connection between LSCF and the alloys through Ag in the paste. If NLSO phase exist at the interface between LSCF and Ag in the paste, the Ag paste much depress ASR owing to low conductivity of NLSO. Various Ag pastes were then mixed with LSCF powder and fused at 800°C. XRD was examined for the fused powders to investigate on the presence of NLSO phase between Ag paste and LSCF. Fig. 4 shows XRD profiles for the fused powders. Crystal phase of $Na_2Si_2O_5$ was observed for 100/0 paste, which was synthesized without SI particles, but no NLSO phase was observed for the fused samples using all the pastes. The Ag pastes contained much amount of Ag and the Ag powder would avoid direct contact between LSCF and glass precursor particles as compared to the flexible sheets composed of NS particles.

The effects of NS/SI ratios of Ag paste on ASR of the interface were then investigated by dc four-probe method. Porous LSCF ceramics were synthesized by sintering at 1330°C and electrical resistance were first measured with various distances between inner electrodes for sensing voltages. The Ag-Si alloy sheets that were coated using Ag paste at various NS/SI ratios were placed between porous LSCF ceramics and fused at 800°C. Electrical resistance for the fused samples was also measured with various distances across the coated Ag-Si alloy sheet. Fig. 5 shows the relationship

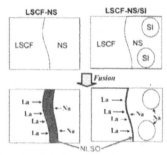

Fig. 3. Schematic illustration of NLSO phase formation at the interface between LSCF and the flexible gas-tight sheet composed of NS and SI particles.

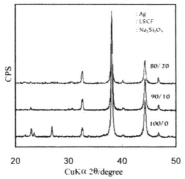

Fig. 4. XRD profiles for the powder fused with LSCF and Ag-containing pastes with various NS/SI ratios.

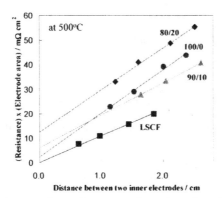

Fig. 5. Relationship between distance between two inner electrodes and (Resistance) x (Electrode area) for LSCF/Ag paste/Ag-Si alloy/Ag pastes/LSCF samples.

between the distance between the inner electrodes and resistances at 500°C. Ordinate axis in Fig. 5 was normalized as (resistance) x (electrode area) to directly estimate ASR at the intercept value of ordinate axis. The intercept for the LSCF sample was almost zero because the LSCF sample had no Ag-Si alloy and Ag paste. The error in this ASR estimation by four-probe method would be small negligible to estimate ASR for the fused samples. The LSCF ceramics fused using Ag-Si alloys coating with various pastes showed different slopes from the LSCF ceramics without the alloy and pastes. Relative densities of LSCF ceramics would be dispersed slightly. Electrical conductivity for porous LSCF is much influenced by porosity and difference in porosity would bring about change in the slopes.

Temperature dependence of the ASR was then estimated using the intercepts in the temperature range of 400°C to 650°C. Fig. 6 shows ASR values for the fused conductive alloys

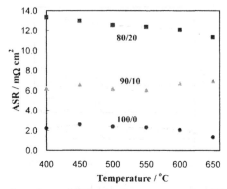

Fig. 6. Temperature dependence of ASR using Ag-pastes at various NS/SI ratios of 100/0. 90/10 and 80/20.

coating with various Ag pastes. ASR was decreased with decreasing SI ratio at all the temperature region. ASR for the fused alloys coating with 80/20 paste was decreased with increasing temperature but that for other alloys showed independent of temperature. The ASR value contains resistances of the alloy and the paste as well as interfacial resistances of LSCF/(Ag paste) and (Ag paste)/(Ag-Si alloy). Conductivities for LSCF. Ag and Ag-Si alloy was almost independent of temperature in the range measured in this investigation. The temperature dependence on ASR for the fused alloys coating with 80/20 paste supposed indirect electrical connection of LSCF-Ag or Ag-alloy. Glassy or crystal phase that showed temperature dependence of the conductivity may be present at the interface of LSCF-Ag or Ag-alloy and led to decrease in conductivity. The fused alloy coating with 100/0 paste exhibited low ASR. Viscosity, wettability of the fusing fluid and reactivity with LSCF or Ag-Si alloy would depend on ASR properties. Ag 100/0 paste would exhibit low viscosity during the fusion as compared to other pastes. Relatively low viscous fusing fluid of 100/0 paste would carry Ag fine particles into LSCF pores and increase contact area between LSCF and Ag. Ag or Ag alloy showed poor wettability on LSCF whereas fusing fluid composed of NS and SI particles showed relatively high wettability on oxide ceramics such as LSCF. Relatively low viscous and highly wettable fusing fluid derived from NS and SI particles would be penetrated into porous LSCF with Ag fine particles and led to increase in ASR.

Fig. 7 shows SEM images of the fused samples using 100/0 paste and 80/20 pastes. Difference in penetration length into porous LSCF was not clearly identified with these SEM images. Both the pastes had tight contact on Ag-Si alloy and porous LSCF. No cracks or delamination was observed in the pastes and at the interfaces.

The effect of alloy composition on resistivities for the fused LSCF ceramic was investigated using Ag 100/0 paste. Ag-Si alloy or Ag-Si-Ge alloy was coated with Ag 100/0 paste. The alloy was placed between porous LSCF ceramics and fused at 800°C. Fig. 8 shows the resistivities at room temperature for LSCF ceramics fused with only Ag-Si-Ge alloy, Ag-Si-Ge alloy coated with 100/0 paste and Ag-Si alloy coated with 100/0 paste. The use of Ag paste led to much decrease in resistivity for the fused LSCF ceramics. and Ag-Si alloy exhibited lower resistivity as compared to Ag-Si-Ge alloy. The use of Ag paste improved ASR on porous LSCF ceramics because relatively low viscous and wettable fusing fluid would carry Ag fine particle and increase in the electrical contact area. The

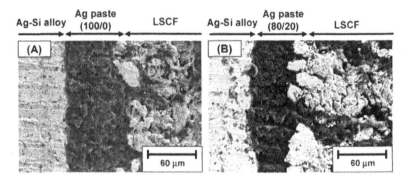

Fig. 7. SEM images of interface of (A) Ag-Si alloy - Ag 100/0paste - LSCF and (B) Ag-Si alloy - Ag 80/20 paste - LSCF after the fusion at 800°C.

use of Ag-Si alloy also improved interfacial resistance between Ag paste and the alloy because Ge-free Ag alloy would prevent the interface between the alloy and fused Ag paste from excess formation of glassy or oxide phase that exhibited low electrical conductivity. Therefore, Ag-Si alloy coating with Ag 100/0 paste resulted in low ASR properties against porous LSCF and this alloy was suitable for the highly conductive gas-tight sealing devices applicable for stacking micro-sized SOFC.

CONCLUSION

We have developed highly conductive gas-tight seals that can melt at temperature lower than 850°C in air and maintain electrical conductive connection between anodes and cathode at 650°C in air as interfacial devices used for stacking micro SOFC. Ag-Si-Ge alloy exhibited highly conductive

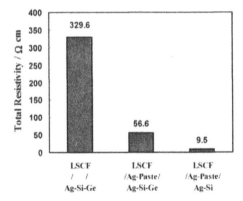

Fig. 8. Comparison of resistivity at room temperature for LSCF/Ag-Si-Ge, LSCF/Ag paste/Ag-Si-Ge and LSCF/Ag paste/Ag-Si.

gas-tight sealing when the alloy was fused on dense ceramics or metal manifolds, but melting Ag-Si-Ge alloy on LSCF porous cathode materials showed relatively high resistance because the Si-Ge region was localized in the vicinity of interfaces against LSCF. Coating on the surface of the alloy was then investigated to avoid direct connection during the fusion. Ag pastes that were composed of fine Ag particles, glass precursor particles and some organic binders were investigated to decrease ASR on porous LSCF. The use of Ag 100/0 paste improved ASR on porous LSCF ceramics because relatively low viscous and wettable fusing fluid would carry Ag fine particle and increase in the electrical contact area. The use of Ag-Si alloy also improved interfacial resistance between Ag paste and the alloy because Ge-free Ag alloy would prevent the interface between the alloy and fused Ag paste from excess formation of glassy or oxide phase that exhibited low electrical conductivity. Therefore, Ag-Si alloy coating with Ag 100/0 paste resulted in low ASR properties against porous LSCF and this alloy was suitable for the highly conductive gas-tight sealing devices applicable for stacking micro-sized SOFC.

ACKNOWLEDGMENTS

This work was supported by NEDO as part of the Advanced Ceramic Reactor Project.

REFERENCES

[1] Y. Fujishiro, M. Awano, T. Suzuki, T. Yamaguchi, K. Arihara, Y. Funahashi and S. Shimizu, Development of the Stacked Micro SOFC Modules using New Approaches of Ceramic Processing Technology, *ECS Trans.*, **7**, 497-501 (2007).
[2] S. Suda, K. Kawahara, K. Jono, M. Matsumiya, Development of Melting Seals Embedded Electrical Conduction Paths for Micro SOFCs, *ECS Trans.*; in received.
[3] S. Suda, K. Kawahara, K. Jono, Development of Insulating and Conductive Seals for Controlled Conduction Paths, *ECS Trans.*, **7**, 2437-2442 (2007).
[4] S. Suda, T. Yoshida, K. Kanamura, T. Umegaki, Formation mechanism of amorphous Na2O-SiO2 spheres prepared by Sol-Gel and Ion-Exchange method, *J. Non-Cryst. Solids*, **321**, 3-9 (2003).
[5] S. Suda, K. Kawahara, K. Jono, Bonding phase of amorphous silicate gas-tight seals for electrode-supported SOFCs, *Trans. Mater. Res. Soc. Japan*, **32**, 111-114 (2007).

Electrolyzer

CARBON DIOXIDE ELECTROLYSIS FOR PRODUCTION OF SYNTHESIS GAS IN SOLID OXIDE ELECTROLYSIS CELLS

Sune Dalgaard Ebbesen and Mogens Mogensen

Fuel Cells and Solid State Chemistry Department, Risø National Laboratory, DTU, Denmark
Frederiksborgvej 399, 4000 Roskilde, Denmark

ABSTRACT
        Carbon dioxide electrolysis was studied in Solid Oxide Electrolysis Cells (SOECs) consisting of a Ni-YSZ support, a Ni-YSZ-cathode layer, and a LSM-YSZ anode as a preliminary investigation for production of synthesis gas. The results of this study show that CO_2 electrolysis is possible in SOECs with nickel electrodes. It was shown that the degradation of the SOEC was around 425 mV pr 1000 hours of operation when operated at 850°C, a CO_2/CO molar ratio of 70/30 and a current density of -0.25 A.cm^{-2}. Increasing the current density to -0.50 A.cm^{-2} increases the degradation of the cell significantly with a degradation rate of 1200 mV pr 1000 hours of operation.

        The degradation rate is around 5 times higher compared to the degradation rate for steam electrolysis in a SOEC at similar conditions. Recent investigations into the degradation mechanism show that the degradation is irreversible when operated at oxidising conditions which indicate that the degradation is not caused by coke formation. The degradation mechanism is currently under further investigation.

INTRODUCTION

        In recent years there has been an increased focus on hydrogen as an alternative energy carrier because of limited fossil fuel sources, increasing oil prices and environmental considerations. Water electrolysis ($H_2O \rightarrow H_2 + \frac{1}{2}O_2$) via solid oxide electrolysis cells (SOEC) for production of hydrogen as alternative energy carrier was under development as an interesting alternative to ordinary alkaline water electrolysis during the early 1980'es [1, 2]. Due to low fossil fuel prices the development of SOEC was slowed down around 1990 but has become increasingly investigated in the recent years as a green energy technology. Beside steam electrolysis, SOECs are also capable of electrolysing carbon dioxide to oxygen and carbon monoxide ($CO_2 \rightarrow \frac{1}{2}O_2 + CO$). This approach was intensively investigated by NASA for production of oxygen [2-6]. CO was treated as an undesired side product and converted into carbon and CO_2 over an iron catalyst, and CO_2 was returned to the SOEC. Both i-V characterization and long-time performance for 1600 hours were reported on nickel based electrodes. Recently i-V characterization for CO_2 electrolysis in "up to date" SOECs was reported which showed significant lower resistance [7]. Also platinum electrodes were applied for the electrolysis of CO_2, although only i-V characterization was reported [8-10].

        Within recent years a huge rise in the number of abnormal weather events has occurred due to a Global Climate Change. Scientists agree that the most likely cause of the changes are man-made emissions of the greenhouse gases. Although there are six major groups of gases that contribute to the global climate change, the most common is carbon dioxide (CO_2). For this reason there are much research in sequestration of CO_2 from power plants and other point sources for storage and removal of CO_2.

        Electrolysis of water and CO_2 in an SOEC yield synthesis gas ($CO + H_2$) which in turn can be catalysed to various types of synthetic fuels [11]. Using SOECs for recycling or reuse of CO_2 from energy systems (or CO_2 capture from air) would therefore be an attractive alternative to storage of CO_2 and would provide CO_2 neutral synthetic hydrocarbon fuels.

        The basic principle for a SOFC and SOEC operating on H_2O/H_2 and CO_2/CO is shown in Figure 1.

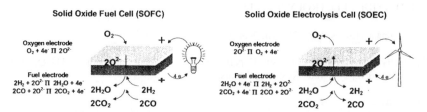

Figure 1. Schematic presentation of the basic operational principle for a Solid Oxide Fuel Cell (SOFC) and a Solid Oxide Electrolysis Cell (SOEC).

For endothermic reactions such as H_2O and/or CO_2 electrolysis it is, from a thermodynamic point of view, advantageous to operate at high temperature as a part of the energy required for water splitting is obtained in the form of high temperature heat e.g. heat form solar concentrators or waste heat from nuclear power plants [12-14]. High temperature electrolysis using SOECs can therefore be operated with a lower electricity consumption compared to electrolysis at low temperature [15, 16]. Furthermore, reaction kinetics increases at high temperature which leads to a decreased internal resistance of the cell and thereby increased efficiency. Even though it is, from a thermodynamic and electrode kinetic point of view, advantageous to operate the SOECs at high temperature, material durability issues makes an upper limit for the operation temperature. Electrolysis in SOECs is typically performed in the temperature range from 750°C to 950°C.

Electrolysis of carbon dioxide may lead to coke formation on the nickel rich fuel electrode as nickel is well known to catalyse dissociation of a carbon containing gases leading to the formation of e.g. carbon nanofibers [17]. Formation of carbon nanofibers are formed on relative large nickel particles only (larger than 5-10 nm). The nickel particles in the SOEC have an average particle size of 1 μm, consequently carbon-fibres are expected to be formed during CO_2 electrolysis. Coke formation is undesirable as carbon may build up in the system. Consequently active sites for electrolysis are reduced. Formation of carbon can originate from carbon monoxide via i.e. via disproportionation, the so-called Boudouard reaction as shown in equation 4. At high temperatures and high CO_2:CO ratios the equilibrium is shifted towards CO which may imply that the reverse Boudouard reaction would occur in the SOEC, consequently coke would not be formed during electrolysis.

$$2CO \rightarrow C + CO_2 \tag{1}$$

The formed carbon (coke) can consist of different types: filamentous (fibres), amorphous and graphitic coke. Filamentous coke contains normally metal particles, whereas amorphous coke contains only little or no metal particles. Graphitic coke is formed mainly at high temperatures (above 900°C) and has a higher density that the other types of coke.

Coke formation may not be critical when operating the SOECs in a mixture of steam and CO_2 as it is well known that increased steam to carbon ratios suppress coke formation during e.g. reforming of methane [18].

Production of Carbon Dioxide Neutral Synthetic Fuels in Solid Oxide Electrolysis Cells

As mentioned above, synthesis gas and thereby synthetic fuels can be produced by electrolysis of carbon dioxide and steam in SOECs. CO_2 capture from air and recycling or reuse of CO_2 from energy systems would therefore be an attractive alternative to storage of CO_2 in the underground (sequestration) and would provide CO_2 neutral synthetic hydrocarbon fuels. The simplest synthetic fuels are methane, methanol and dimethyl ether.

Methane is typically produced from carbon monoxide and hydrogen over nickel based catalysts, at temperatures between 200 °C and 450 °C. Pipe lines and storage facilities for natural gas are widely established in Denmark. In this perspective, it is interesting to produce methane from renewable energy sources to feed in to the natural gas pipe lines. Since the negative electrode of a SOC is partly made of Ni it is in principle possible to produce CH_4 within the cell (at high pressure and low temperature) [19]. The entropy change for CH_4 production from CO_2 and H_2O is nearly zero. This means that the overall efficiency for a conversion of electricity to CH_4 and back again can be very high, if the reaction kinetics are fast, since only small reaction entropy losses occur. Also methanol and dimethyl ether may be produced from synthesis gas over cupper based catalysts operated at temperatures between 200 °C and 300 °C and at a pressure of 45 – 60 bar [20, 21].

Capture of CO_2 for recycling can be achieved by absorption processes employing amines or carbonates as absorbents. The regeneration includes heating of the absorbent; therefore reduction of the energy requirement becomes a determining factor for realizing CO_2 recycling. From the viewpoint of energy saving in regeneration of the absorbent, carbonates are preferable to amine solutions, since the energy requirement for CO_2 removal in the carbonate process is about half of that of the amine process [22]. However the rate for CO_2 absorption and desorption with carbonates is slow, but for CO_2 capture/recycling from air, the absorption and desorption rate may not be a determining factor. Mineral carbonation has been recognized as a potentially promising route for permanent and safe storage of carbon dioxide, and thereby also a promising route for recycling of CO_2. A number of different carbonation processes has been reported, of which aqueous mineral carbonation route was selected as the most promising in a recent review [23].

Calcium hydroxide is a well known CO_2 absorbent [24], whereas also the less known magnesium hydroxide or calcium magnesium hydroxide can be employed. Thermodynamic equilibrium calculations (calculated using Factsage 5.5 Software [25] at 1 atm) have been performed to determine the energy consumption necessary for the regeneration. CO_2 capture/recycling using magnesium hydroxide/carbonate can be operated at approximately 400 °C lower than the case for calcium hydroxide. A carbon neutral energy cycle utilizing CO_2 capture from air with magnesium hydroxide/carbonate in combination with a SOFC and SOEC is sketched in Figure 2. Magnesium carbonate is abundant in nature as calcium magnesium carbonate. A carbonate cycle for CO_2 capture with calcium magnesium carbonate can be operated between 250 °C and 400 °C utilizing magnesium carbonate only. Using only magnesium carbonate from calcium magnesium carbonate, higher amount of minerals would have to be mined and transported.

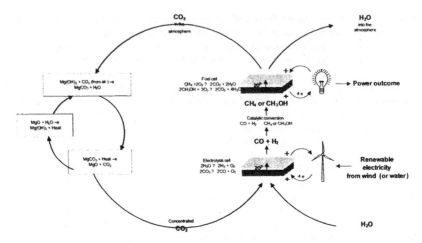

Figure 2. Carbon dioxide neutral energy cycle utilizing CO_2 capture from air with magnesium carbonate in combination with a SOEC, after [26].

A hydroxide/carbonate cycle for CO_2 capture and recycling in combination with an SOEC for production of synthetic fuel is definitively technically feasible, but the practical and economic aspects regarding capture/recycling and the electrolysis parameter have to be assessed to determine the most suitable methods for CO_2 capture/recycling and electrolysis operation. It is the aim of this paper to examine the SOEC feasibility for CO_2 electrolysis. SOEC performance and durability over many hundred hours is here examined in CO_2/CO mixtures without water to separately examine the mechanism of CO_2 electrolysis as an initial step for production of synthetic fuel in SOECs.

EXPERIMENTAL

Cell characterization and electrolysis measurements
      Planar cells of 5×5 cm^2 Ni/YSZ supported DK-SOFC cells with an active electrode area of 4×4 cm^2 were used for the electrolysis tests. The cells are full cells produced at Risoe National Laboratory [27, 28]. Detailed description of the setup and the start-up procedure i.e. heating up and reduction of NiO to Ni in flowing hydrogen is given elsewhere [29, 30].
      After reduction the hydrogen flow was slowly stepped to zero while first CO_2 was added followed by the addition of CO (to avoid initial carbon deposit) to give CO_2/CO molar ratio of 50/50 with a flow rate of 23 L/h while the gas passing over the oxygen electrode was changed from air to 20 L/h of O_2. DC characterization of the cell was performed by recording polarization curves (i-V curves) in both electrolysis mode and fuel cell mode by controlling the current. Subsequently the CO_2/CO ratio was increased to 70/30 and additional DC characterization was performed.
      After the initial characterisation the cell was switched from OCV to electrolysis operation at a CO_2/CO ratio of 70/30 and current density of -0.25 A.cm^{-2} or -0.50 A.cm^{-2}. Both at OCV and during electrolysis the resistance over the cell was analysed by electrochemical impedance spectroscopy (EIS) in combination with an external shunt to measure the AC current through the cell (Solartron 1255 or Solartron 1260 frequency analyzer).

RESULTS AND DISCUSSION

The initial performance of the cell was measured by recording i-V curves at 850°C in both CO_2/CO and H_2O/H_2 mixtures (CO_2/CO molar ratio of 50/50 and 70/30 and H_2O/H_2 molar ratio of 50/50). Figure 3 shows a comparison of initial i-V curves measured in CO_2/CO mixtures with i-V curves measured in H_2O/H_2 mixtures.

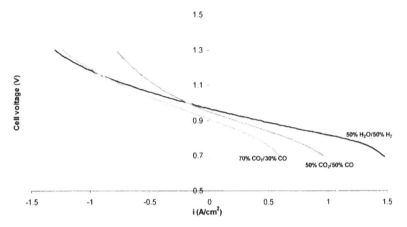

Figure 3. Initial DC characterization for CO_2/CO mixture (ratio = 50/50 and 70/30) and H_2O/H_2 mixture (ratio = 50/50) at 850°C.

From the i-V characteristic shown in Figure 3, it is observed that no discontinuity occurs in the shift from fuel cell to electrolysis operation. The cell had an open-circuit voltage (OCV) of 967 and 948 mV at 850°C for steam electrolysis (50% H_2O, 50% H_2) and carbon dioxide (50% CO_2, 50% CO) respectively. The higher open-circuit voltage for CO_2 electrolysis is in agreement with the higher Nernst potential for CO_2 electrolysis compared to steam electrolysis at 850°C. Increasing the carbon dioxide concentration (70% CO_2, 30% CO) decreases the OCV to 912 mV, again in agreement with the Nernst potential.

Calculating the area specific resistance (ASR) as the chord from OCV to the voltage measured at a current density of 0.20 $A.cm^{-2}$ (fuel cell mode) for the i-V curve leads to ASR at 0.17 $\Omega.cm^2$ for the cell when characterised in 50% H_2O / 50% H_2 at 850°C in agreement literature [31]. Calculating the ASR in electrolysis mode (from OCV to the voltage measured at a current density of -0.20 $A.cm^{-2}$) leads to ASR at 0.18 $\Omega.cm^2$, showing a higher activity towards oxidation of H_2 than reduction of H_2O for the Ni/YSZ electrode as found in literature [29, 32].

The ASR for oxidation of CO, when operated in CO_2/CO mixtures with a ratio of 50/50 and 70/30 was 0.22 and 0.23 $\Omega.cm^2$ respectively, showing lower activity for CO oxidation compared to H_2 oxidation as previously reported for Ni/YSZ electrodes [33]. For reduction of CO_2 (CO_2 electrolysis), ASR of 0.26 and 0.29 $\Omega.cm^2$ was found for reduction in CO_2/CO mixtures with a ratio of 50/50 and 70/30 respectively. Consequently, similar to oxidation and reduction of H_2O/H_2, a higher activity for oxidation of CO compared to reduction of CO_2 was found.

After testing the initial performance of the cell, durability test at constant galvanostatic electrolysis conditions were performed at a CO_2/CO ratio of 70/30 and a current density of either -0.25 $A.cm^{-2}$ (Figure 4) or -0.50 $A.cm^{-2}$ (not shown).

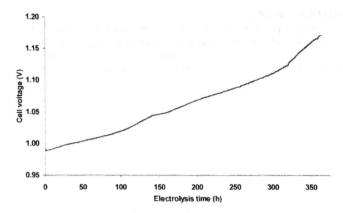

Figure 4. Cell voltage measured during electrolysis test at -0.25 A.cm^{-2}. The gas composition to the negative fuel electrode was a mixture of CO_2 and CO (CO_2/CO ratio of 70/30) while O_2 was passed over the oxygen electrode.

From Figure 4 it can be seen that the increase in cell voltage is close to linear, which may indicate coke formation on the Ni/YSZ electrode. After 370 hours of electrolysis the test was stopped because of a failure in the gas supply that damaged the cell. Consequently, no reliable post mortem analysis of the cell is reported. The increase in cell voltage during the electrolysis test corresponds to a degradation rate of 425 mV pr. 1000 hours of operation (Figure 4). Increasing the current density of -0.50 A/cm^2 (not shown) increases the degradation of the cell significantly with a degradation rate of 1200 mV pr 1000 hours of operation. The degradation rate for CO_2 electrolysis is around 5 times higher compared to the degradation rate of steam electrolysis at similar conditions SOECs [29].

Electrochemical impedance spectra were recorded during the electrolysis test to examine the cause of the degradation. Nyquist plot of the resulting impedance spectra are shown in Figure 5.

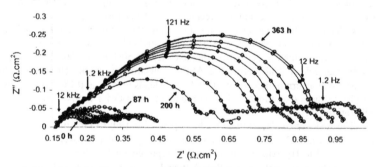

Figure 5. Nyquist plot of impedance spectra obtained during CO_2 electrolysis test at -0.25 A.cm^{-2}. The gas composition to the fuel electrode was a mixture of CO_2 and CO (CO_2/CO ratio of 70/30) while O_2 was passed over the oxygen electrode. Frequencies are given for the closed symbols.

The initial impedance spectra during electrolysis consists of two arcs with summits of 1.2 Hz and 2.6 kHz which can be assigned to gas conversion (0.1 – 10 Hz) [34] and gas-solid (desorption, absorption, dissociation) or solid-solid (surface diffusion, double layer reaction) (1 kHz – 50 kHz) [34] respectively. After around 30 hours of electrolysis a third arc start to develop at 100Hz, which may be assigned to changes in diffusion (10 Hz – 1 kHz) [35].

To examine the changes in the impedance spectra during electrolysis, the difference in the resistance as a function of time is calculated as $\Delta_t \partial Z'$ [36]. The resulting spectra enable determination of the processes causing the altered resistance and are shown in Figure 6.

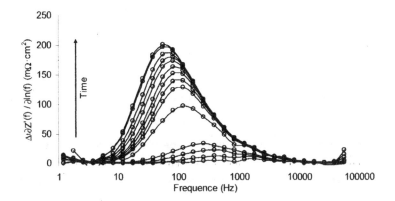

Figure 6. $\Delta_t \partial Z'$ during CO_2 electrolysis at -0.25 A.cm^{-2}. The gas composition to the fuel electrode was a mixture of CO_2 and CO (CO_2/CO = 70/30) while O_2 was passed over the oxygen electrode.

The differential impedance spectra during the electrolysis show that the main change occur between 60-80 Hz, which may be assigned to changes in diffusion [35]. Increased diffusion limitations may be caused by the formation of coke in the porous hydrogen electrode. Recent experimental evidence show that the degradation is irreversible when the cell is set to OCV or when the cell is operated in fuel cell mode (oxidising conditions) [37]. The irreversible degradation indicates that the degradation is not caused by coke formation. On the other hand other interactions between CO (carbon) and the nickel particles such cannot be excluded at present.

If this technology should play an important role in the future it is necessary to lower the degradation rate and to find cheap ways to collect CO_2 from the air or from other sources. The degradation mechanism is currently under further investigation by varying the electrolysis conditions (current, temperature and CO_2/CO ratio); by thermo gravimetric analysis as well as electron microscopy on both durability tested and reference cells.

CONCLUSION

SOFCs were operated with CO_2 electrolysis for several hundreds of hours 850°C. At a CO_2/CO molar ratio of 70/30 and a current density of -0.25 A.cm^{-2} the cell voltage increased from 990 mV to 1171 mV during 370 hours of electrolysis which corresponds to a degradation rate of 425 mV pr 1000 hours of operation. Increasing the current density to -0.50 A/cm^2 increases the degradation of the cell significantly with a degradation rate of 1200 mV pr 1000 hours of operation.

The degradation rate is around 5 times higher compared to the degradation rate of steam electrolysis in a SOEC. The degradation during CO_2 electrolysis is irreversible when operated at oxidising conditions which indicate that the degradation is not caused by coke formation. Other interactions between CO (carbon) and the nickel particles cannot be excluded at present and the degradation mechanism is currently under further investigation.

ACKNOWLEDGEMENTS
This work was supported financially by the Danish National Advanced Technology Foundation through the Advanced Technology Platform on 2nd Generation Bio Ethanol. The support from the staff of the Fuel Cells and Solid State Chemistry Department, Risø-DTU, is highly appreciated.

REFERENCES
[1] W. Doenitz, R. Schmidberger, E. Steinheil, R. Streicher, Int. J. Hydrogen Energy 5 (1980) 55-63.

[2] A.O. Isenberg, Solid State Ionics 3-4 (1981) 431-437.

[3] L. Elikan, J.P. Morris, C.K. Wu, NASA Research Center NASA Report CR-2014 (1972)

[4] A.O. Isenberg, NASA Research Center NASA Report CR-185612 (1989)

[5] A. O. Isenberg and C. E. Verostko, in: 19th Intersociety Conference on Environmental Systems, ed. San Diego, California, USA, 1989) Paper 891506-

[6] A. O. Isenberg and R. J. Cusick, in: 18th Intersociety Conference on Environmental Systems, ed. San Franscisco, California, USA, 1989) Paper 881040-

[7] S.H. Jensen, P.H. Larsen, M. Mogensen, Int. J. Hydrogen Energy 32 (2007) 3253-3257.

[8] K.R. Sridhar , B.T. Vaniman, Solid State Ionics 93 (1997) 321-328.

[9] G. Tao, K.R. Sridhar, C.L. Chan, Solid State Ionics 175 (2004) 621-624.

[10] G. Tao, K.R. Sridhar, C.L. Chan, Solid State Ionics 175 (2004) 615-619.

[11] I. Chorkendorff and J. W. Niemantsverdriet, Concepts of Modern Catalysis and Kinetics, Wiley-VCH, Weinheim (2006).

[12] B. Yidiz , M.S. Kazimi, Int. J. Hydrogen Energy 31 (2006) 77-92.

[13] J. E. O'Brien, C. M. Stoots, J. S. Herring, P. A. Lessing, J. Hartvigsen, and S. Elangovan, in: Proceedings of ICONE12, 12th International Conference on Nuclear Engineering, ed. Virginia, USA, 2004)

[14] A. Hauch, S. H. Jensen, M. Menon, and M. Mogensen, in: Proceedings of Risø International Energy Conference, Denmark, ed. S. L. Petersen and H. Larsen (2005) 216-230.

[15] M.Mogensen, S.H.Jensen, A.Hauch, I.Chorkendorff, and T.Jacobsen, in: Proceedings 7th European SOFC Forum, ed. Lucerne, 2006)

[16] S. H. Jensen, Solid oxide electrolyser cell, Ph.D. thesis, Risø National Laboratory, Denmark (2006).

[17] K.P. de Jong , J.W. Geus, Catal. Rev-Sci. Eng 42 (2000) 481-510.

[18] J.P. Vanhook, Catal. Rev-Sci. Eng 21 (1980) 1-51.

[19] S. H. Jensen, J. V. T. Høgh, R. Barfod, and M. Mogensen, in: Proceedings of Risø International Energy Conference, Denmark, ed. S. L. Petersen and H. Larsen (2003) 204-207.

[20] G.A. Mills, Fuel 73 (1994) 1243-1279.

[21] J.B. Hansen, Handbook of Catalysis,Wiley-VCH, New York, 1997.

[22] G. Astarita, D. W. Savage, and A. Bishio, Gas Treating with Chemical Solvents, John Wiley & Sons (1983).

[23] W.J.J. Huijgen, G.J. Ruijg, R.N.J. Comans, G.J. Witkamp, Ind. Eng. Chem. Res. 45 (2006) 9184-9194.

[24] W.J.J.Huijgen and R.N.J.Comans, Carbon dioxide storage by mineral carbonation, Energy Research Centre of The Netherlands, Cheltenham, United Kingdom (2005).

[25] C.W.Bale, P.Chartrand, S.A.Degterov, G.Eriksson, K.Hack, R.Ben Mahfoud, J.Melançon, A.D.Pelton, and S.Petersen, FactSage Thermochemical Software and Databases, CRCT - Center for Research in Computational Thermochemistry École Polytechnique (Université de Montréal), Box 6079, Station Downtown, Montréal, Québec, CANADA and GTT-Technologies, Kaiserstrasse 100, 52134 Herzogenrath, GERMANY (2002).

[26] A. Hauch, S. H. Jensen, S. D. Ebbesen, and M. Mogensen, in: Proceedings of Risø International Energy Conference, Denmark, ed. S. L. Petersen and H. Larsen (Roskilde, Denmark, 2007) 327-338.

[27] A. Hagen, R. Barfod, P.V. Hendriksen, Y.L. Liu, S. Ramousse, Journal of the Electrochemical Society 153 (2006) A1165-A1171.

[28] P. H. Larsen, C. Bagger, S. Linderoth, M. Mogensen, S. Primdahl, M. J. Jorgensen, P. V. Hendriksen, B. Kindl, N. Bonanos, F. W. Poulsen, and K. A. Maegaard, in: Proceedings - Electrochemical Society, ed. H. Yokokawa and Singhal S.C. (2001) 28-37.

[29] A. Hauch, S.H. Jensen, S. Ramousse, M. Mogensen, J. Electrochem. Soc. 153 (2006) A1741-A1747.

[30] M. Mogensen ,P.V. Hendriksen, High temperature solid oxide fuel cells - Fundamentals, design and applications, Chapter 10, Elsevier, London, 2003.

[31] N. Christiansen, J.B. Hansen, H. Holm-Larsen, S. Linderoth, P.H. Larsen, P.V. Hendriksen, A. Hagen, Proceedings of Risø International Energy Conference, Denmark (2007)

[32] O.A. Marina, L.R. Pederson, M.C. Williams, G.W. Coffey, K.D. Meinhardt, C.D. Nguyen, E.C. Thomsen, J. Electrochem. Soc. 154 (2007) B452-B459.

[33] P. Holtappels, L.G.J. de Haart, U. Stimming, I.C. Vinke, M. Mogensen, J. Appl. Electrochem. 29 (1999) 561-568.

[34] S. Primdahl , M. Mogensen, J. Electrochem. Soc. 145 (1998) 2431-2438.

[35] S. Primdahl , M. Mogensen, J. Electrochem. Soc. 146 (1999) 2827-2833.

[36] S.H. Jensen, A. Hauch, P.V. Hendriksen, M. Mogensen, N. Bonanos, T. Jacobsen, J. Electrochem. Soc. 154 (2007) B1325-B1330.

[37] S.D. Ebbesen, R. Knibbe, M. Mogensen, To be published (2008)

Author Index

Printed and bound by CPI Group (UK) Ltd, Croydon, CR0 4YY

16/04/2025

14658452-0004